Lecture Notes in Networks and Systems

Volume 208

The series "Lecture Notes in Networks and Systems" publishes the latest developments in Networks and Systems—quickly, informally and with high quality. Original research reported in proceedings and post-proceedings represents the core of LNNS.

Volumes published in LNNS embrace all aspects and subfields of, as well as new challenges in, Networks and Systems.

The series contains proceedings and edited volumes in systems and networks, spanning the areas of Cyber-Physical Systems, Autonomous Systems, Sensor Networks, Control Systems, Energy Systems, Automotive Systems, Biological Systems, Vehicular Networking and Connected Vehicles, Aerospace Systems, Automation, Manufacturing, Smart Grids, Nonlinear Systems, Power Systems, Robotics, Social Systems, Economic Systems and other. Of particular value to both the contributors and the readership are the short publication timeframe and the world-wide distribution and exposure which enable both a wide and rapid dissemination of research output.

The series covers the theory, applications, and perspectives on the state of the art and future developments relevant to systems and networks, decision making, control, complex processes and related areas, as embedded in the fields of interdisciplinary and applied sciences, engineering, computer science, physics, economics, social, and life sciences, as well as the paradigms and methodologies behind them.

Indexed by SCOPUS, INSPEC, WTI Frankfurt eG, zbMATH, SCImago.

All books published in the series are submitted for consideration in Web of Science.

More information about this series at http://www.springer.com/series/15179

Grzegorz Sierpiński · Elżbieta Macioszek
Editors

Decision Support Methods in Modern Transportation Systems and Networks

Springer

Editors
Grzegorz Sierpiński
Faculty of Transport and Aviation Engineering
Silesian University of Technology
Gliwice, Poland

Elżbieta Macioszek
Faculty of Transport and Aviation Engineering
Silesian University of Technology
Gliwice, Poland

ISSN 2367-3370 ISSN 2367-3389 (electronic)
Lecture Notes in Networks and Systems
ISBN 978-3-030-71770-4 ISBN 978-3-030-71771-1 (eBook)
https://doi.org/10.1007/978-3-030-71771-1

This Springer imprint is published by the registered company Springer Nature Switzerland AG
The registered company address is: Gewerbestrasse 11, 6330 Cham, Switzerland

Preface

The structure, efficiency and functionality of future transport systems depend to a large degree on the decisions being made. With the passing of time, the existing solutions are becoming insufficient on account of the increasing mobility, while the cargo transport-related needs constantly grow. Another problem which must be taken into consideration as an aspect of decision-making problems is the necessity of reducing the negative environmental impact of transport. The contemporary solutions require networks and systems that have been adequately prepared. Hence the immense importance of support for research methods.

With regard to the foregoing, the main goal of the presented monograph entitled *Decision support methods in modern transportation systems and networks* is to demonstrate and discuss the issues related to the research methods used to support solving of decision-making problems in the contemporary transport networks and systems. This monograph is a detailed treaty on such specific subjects as the research methods pertaining to both the existing and future networks and systems, with special emphasis on the passenger and cargo transport solutions. In numerous cases, the support may be sought in the application of multi-criteria decision-making methods, but the monograph elaborates upon far more alternative solutions. Particular attention has been attached to environmentally friendly solutions requiring transport networks to be redesigned or clean zones to be implemented. Both with regard to passenger transfers and freight, there is a clear need for solutions which integrate network streams into transport chains. What becomes very significant when addressing the problems of passenger transport by applying a system approach is adequate planning of shared mobility accompanied by optimisation of the public transport network. Considering cargo transport, on the other hand, for the sake of its efficiency, it is extremely important to apply the synergistic approach and to address the last mile problem in an optimised manner by taking into account diverse methods for siting of logistics centres in transport networks.

The selection of methods described in the monograph can be used as the supporting means to solve decision-making problems, having a direct impact on the nature of the smart transportation systems of the future.

Katowice, Poland
January 2021

Grzegorz Sierpiński
Elżbieta Macioszek

Contents

The Redesign Methodology of a Transportation Network

Remigiusz Wiedemann and Hanna Sawicka

1 Introduction

The analysis and redesign of the existing land transportation network is usually a rough task, because of the complex network structure and the scope of the potential changes. This network could be considered as a set of the following elements:

- infrastructure, including streets represented by arcs of the transportation network, intersections performed by nodes of the transportation network, stops, traffic lights location;
- organizational and safety rules, including: traffic, priority rules at collision points;
- types of transportation means allowed for traffic e.g. cars, buses, trams;
- human behavior represented by the percentage of travellers using a particular type of a transportation mean i.e. modal split.

All those components of the transportation network are related to each other and they create the system. Its analysis aiming at evaluation and introduction of potential changes can be carried out in two ways [1], i.e. directly on the real network or on its model. The first approach is rarely cost-effective and it is usually connected with disruption of the system. Moreover, it can be connected with the risk in case the implementation of the new solution doesn't meet the customers' needs. In such circumstances the second way of analysis is advised. This is the experiment with the model, which represents the system. The first type of the experiment, which could be carried out is on iconic model, such as vehicle simulator. However, most of the systems, including transportation networks, are represented by mathematical

R. Wiedemann · H. Sawicka (✉)
Poznan University of Technology, Institute of Transport, Poznan, Poland
e-mail: hanna.sawicka@put.poznan.pl

R. Wiedemann
e-mail: remigiusz.wiedemann@doctorate.put.poznan.pl

G. Sierpiński and E. Macioszek (eds.), *Decision Support Methods in Modern Transportation Systems and Networks*, Lecture Notes in Networks and Systems 208,
https://doi.org/10.1007/978-3-030-71771-1_1

models. This is the second type of the experiment with the model of the system. It has the quantitative character and it is usually connected with the *what if* analysis. To get the exact solution of this analysis the computational experiments are carried out. Depending on the complexity of the modeled situation, different computing resources are required. Law and Kelton [1] state that if the analytical solution of a mathematical model is available and it is computationally efficient, it is desirable to study the model this way. Otherwise, the model is analyzed by means of simulation. There are three dimensions of simulation models classification, such as [1]:

- static or dynamic,
- deterministic or stochastic,
- continuous or discrete.

This classification reflects the importance of time, randomness of components and changes of a system at countable (or not) points in time. In static simulations time plays no role, whereas dynamic simulation models reflect the system evolving over time. The model without the probabilistic values is deterministic, while the model with random input numbers is stochastic. Continuous simulation models represent systems with the state that changes continuously over time. In discrete simulation models at least one phenomena of interest change value or state at discrete points in time [2].

The analysis of the current state of the transportation network usually leads to a conclusion that changes are necessary. They are obtained through the redesign process in the structure of the transportation network, including an introduction of new arcs and nodes; new traffic lights location; determination of the number, type and location of vehicle detectors based on sensor technologies e.g. inductive loop detectors; signal timing; setting the logic of intelligent control, sometimes also charging stations sitting [3], special routes for freight transport [4] and functionalities for autonomous cars [5].

The redesign of the complex transportation network presented in the literature can be divided into two major methodological approaches, i.e. an analytical solution of the problem [3, 7, 8] and a simulation [1, 9, 11]. To respond to the analytical solution Leblanc [7] presents nonlinear mixed integer programming model for the discrete network design problem. This model is verified on the Sioux Falls city network, composed of 24 nodes. The problem of urban road network improvement aiming at minimization of total congestion is solved by a sequence of shortest route problems. In contrast to a single criterion evaluation, Mizuno et al. [8] present the multi-objective optimization approach, with discrete decision variables. The authors solve the problem of growing network by the addition of a node in order to obtain the desired network. They use different criteria, like an average path length and a clustering coefficient, which can be applied, when the desired network structure is going to be designed. The solution of the problem can be obtained with a use of multi objective genetic algorithm. The other analytical approach, which is based on a simulated annealing algorithm, is applied by Friesz et al. [6]. The authors present the solution of the continuous network problem, where globally optimal solution is found. The procedure proposed by the authors is verified on two networks, i.e. 16 arcs and 16 paths with different travel demand scenarios.

All cases show that the analytical approach provides the optimal solution of the problem, which is its main advantage. It is characterized by the precise and exact information. However, most of the analytical solutions of the transportation network design and redesign are static. They do not reflect the dynamic character of the problem. Moreover, they are usually based on deterministic values of the modeled network, while its nature is nondeterministic.

The dynamically changing operations in the transportation network and non-deterministic information can be modeled with an application of simulation. This approach is close to the real and continuously changing situations. This fact emphasize Abid and Hussain [9] by presenting the application of simulation in transportation network planning. The authors model future origin-destination matrices for the morning and evening peak hours in Al-Mansour city, Iraq. They concentrate on traffic volumes, building a model and implementing it in TransCad software. As a result the trips are allocated to the routes and the network is evaluated. The simulation approach extended by a multiple criteria analysis is presented by Abdelghany and Mahmassani [10]. The authors show a simulation model for an urban transportation network. Its main issue is the relation between the mode choice (buses, subway, high-occupancy vehicles and private cars) and traffic assignment. The travelers' mode and route selection has a multiple criteria character. The authors point out the dynamic interaction between mode choice and traffic assignment as well as the importance of proposed mutliobjective assignment procedure. The problem of dynamically changing transportation network is also the subject of the research analysis carried out by Bloomberg and Dale [11]. The authors compare two very common traffic simulation tools, which are Vissim and Corsim. L. Bloomberg and J. Dale build the simulation models of the congested transportation network in both tools and they carry out the simulation experiments. The obtained results are consistent, which makes their analysis more reliable, with high confidence level.

The approaches presented in the literature are mostly concentrated on the part of the transportation network redesign, e.g. trips' allocation to the routes, mode choice. The solutions have either global or local character. Part of them is leading to the optimal solution, while the others are approaching the best solution.

The research analyzes are usually single-criterion. However, the character of network problems is complex, where different and contradictory points of view should be taken into account. In such circumstances the multicriteria decision aiding (MCDA) [12–15] approach can be applied. It helps the decision maker solving the problem, presenting the compromise solution. As P. Vincke [14] states MCDA is a field which aims at giving the decision maker (DM) some tools in order to enable him/her to solve a complex decision problem where several points of view must be taken into account. Belton and Stewart [16] define MCDA as a collection of formal approaches, which seek to take explicit account of multiple criteria in helping individuals or groups explore decisions that matter. In such decisions the consequences are substantial, impacts are long-term and may affect many people, and mistakes might not easily be remedied [16]. The decisions are based on a compromise, which is a trade-off between criteria and the DM's preferences. The solution is selected from the family of variants, also called alternatives or scenarios. They can be constructed in different ways, i.e. two or more alternatives can be introduced conjointly as a final solution or it is assumed that the variants are totally independent and they cannot be implemented together.

The variants evaluated by the family of criteria have the following features [12]:

- completeness due to the decision-making aspects of the considered problem;
- appropriate formation, taking into account the global preferences of the DM;
- non-redundancy, i.e. a situation in which semantic ranges of criteria are not repeated.

The performances of variants evaluated by the set of criteria are usually expressed as deterministic values. However, the assessment of real-world situations modeled in a simulation tool becomes more complex, because their performances are expressed by non-deterministic information. MCDA methods applied to support the DM are mostly based on stochastic approach for modeling their preferences. Based on the authors' experience the methods dedicated to solve complex decision transportation problems with non-deterministic criteria values and DM's preferences are not efficient enough.

Thus, the aim of this chapter is to present the methodology of transportation network redesign, which includes macroscopic and microscopic approach, including traffic signal controls as well as stochastic multiple criteria decision aiding approach.

The macroscopic phase of the methodology provides the overall analysis, with global information about the transportation network. The microscopic phase specifies the selected part of the analyzed network in details. The simulation tool, reflecting the dynamic character of the transportation network, supports the methodology. The evaluation of different redesign scenarios with non-deterministic information about their performances is carried out with the application of the stochastic MCDA methodology, including the selection of the most suitable MCDA method, computational experiments with stochastic ranking construction, assessment of the final solution, resulting in a recommended solution.

The chapter is composed of five sections. The first one presents the introduction to the analysis of the transportation network. The main approaches are presented and the literature review is carried out. The second section is devoted to different approaches to the land transportation network redesign process supported by the literature review. In the third section, there is presented the methodology proposed by the authors of this chapter. Its major phases and steps are described in details. The fifth section presents the verification of the proposed approach on the real-world example. The current state of the analyzed transportation network with the decision problem is described, as well as redesign scenarios of this network. The set of 5 variants, modeled in the simulation tools, is evaluated by the family of 6 criteria. The computational experiments, leading to the ranking of variants and compromise solution recommendation, are carried out with an application of the selected ranking MCDA method supported by the classification approach [17]. Finally, in the fifth section, conclusions are drawn. The chapter is completed with the list of references.

2 Redesign of Transportation Networks

2.1 Determinants for the Choice of Traffic Intersection

The traffic organization in the urban area, which is characterized by high complexity, requires careful analysis. Defining the scope of planned modifications in the existing transportation system is a crucial aspect of traffic redesign process.

On some occasions, there is no necessity to redesign the communication infrastructure comprehensively in order to improve traffic conditions. Sometimes, the improvement of traffic conditions can be achieved by adjusting the signposting at a critical intersection. However, far more expensive investments are considered in times of increasing urban congestion, especially during peak hours. These investments include a complex redesign of intersections, an implementation of traffic light management as well as development and promotion of public transportation.

In case of a comprehensive reorganization at the intersection of at least two roads, the selection of the appropriate type of intersection depends on many factors, such as the required capacity, safety considerations, restrictions related to areas for investments, financial considerations, importance of public communication or the possibility of further expansion [18, 19].

Most often, the selection of type of intersection in the urban area is limited to the following types of junctions:

- interchanges,
- priority to the right intersections,
- roundabout,
- signalized intersections.

An interchange may be a perfect solution providing the highest level of capacity and safety at very high daily volume of traffic. It proves efficient both in major and minor road due to the separation of individual collision flows. However, this is a highly expensive investment and usually requires a lot of space for its implementation. Therefore, interchanges are usually built on the outskirts of cities combining crucial local, expressways, beltways, etc. [20–22].

Priority to the right intersections does not seem to be the best option in case of intensive road congestion due to low level of safety on turning relations and difficulties with entering the traffic by the drivers at minor inlets [21].

A relatively more difficult decision-making case is the choice between traffic lights and a roundabout, especially when significant intensity is expected at all intersections. The decision should be taken on the basis of the following factors: the speed limit, a number of public-service passenger vehicles in traffic, pedestrian and bicycle traffic volume, available space, number of road accidents and a green wave [21]. A similar case was considered by R. Wiedemann [22] in the area of the planned construction of a five-leg intersection concentrating high flows of road traffic and public transport vehicles. The final recommendation of adaptive, coordinated signaling on channelized intersection allowed to obtain a significant improvement in traffic conditions in relation to the existing state.

Generally, an implementation of a roundabout immediately leads to significant delays of vehicles in main traffic flow. An unquestionable advantage of roundabouts is a very small number of conflict points but this only applies to single-lane roundabouts. The increase in the number of conflict points is inversely proportional to the level of security at the intersection [20, 21, 23].

The installation of traffic lights is a competitive solution for the construction of roundabouts. It increases delays in the main traffic flow due to the queue of vehicles standing at a red light. However, the effects of this phenomenon can be minimized by using intelligent traffic control systems and coordination of control programs. The high number of conflicts is reduced by allocating particular relations to appropriate phases which allow a safe traffic flow.

The proposed approach allows for the initial selection of possible solutions. Nowadays, majority of solutions can be tested before implementation.

2.2 *Applications Developed for Transportation Network Redesign*

In order to increase the efficiency of the design of new solutions in the field of traffic engineering advanced computer tools are currently used, which according to the classification adopted by the authors can be divided into:

- traffic engineering softwares,
- advanced simulation programs.

The first group includes the LISA+ software developed by Schlothauer and Wauer [24]. The main functionality of Lisa+ is based on designing and testing the fixed and adaptive traffic light systems. Moreover, in cooperation with PTV Vissim it may be used to a comprehensive traffic simulations in microscopic scale [24–27].

The second group is composed of tools used to simulate transportation on various levels of detail. The type of the model reflects their complexity. Therefore, these tools and models can be divided into three types [28, 29]:

- macroscopic,
- microscopic,
- mesoscopic.

Macroscopic tools and models are used for simulating the traffic on the general level of detail. The basic characteristics include average velocity, traffic density, traffic volume, etc. These models support planning large transportation networks for the needs of, for example, cities or voivodships. Macroscopic models allow for predicting how specific investments can affect traffic in a particular research area.

One of the most popular tool using the assumptions of this model is the product of PTV Group, Visum [26, 28, 29].

Macroscopic models have significant influence on road infrastructure planning. In fact, most of big cities has their own simulation model these days. Municipalities usually make it available to external entity for the needs of planned road traffic management projects [29].

In fact these type of modeling represents a low level of detail in the movement parameters. It means the interaction between individual flow participants are not observed. Final results of the simulation are highly aggregated. Therefore, when the traffic analysis requires consideration of such characteristic as average queue length on intersection inlets, number of vehicle stops or fuel consumption, macroscopic model becomes inadequate. Then, it is necessary to use microscale modeling. This type of approach focuses on the mathematical description of the movement of each individual vehicle or pedestrian in simulation experiment. It has practical use in tests of traffic lights systems. The prominent application using this type of approach is PTV Vissim [26, 29].

This application allows to simulate movement on different road infrastructure and traffic management, especially on intersections as a place where comes to the highest number of interactions between traffic participants. The movement imitation is based on a car-following model developed by Wiedemann [30] and a line changing model. The Vissim software became a widely popular tool due to its simplicity of use and usefulness in testing variable systems of traffic light systems (from fixed-time to accommodative and coordinated across multiple nodes signaling programs).

The mesoscopic approach, which combines the features of micro and macro scale modelling is the last method used for imitating movement on road infrastructure. The characteristics describing the movement are less aggregated than in the macroscopic models, but the interactions between individual vehicles are analyzed at more general level than the micro scale models. For practical uses this type of modeling is applied in CONTRAM software [28].

Since there are many simulation tools helping the modeling process of the transportation network redesign concepts, the problem is the selection of the best solution. Taking into consideration the complexity of the considered issue, the MCDA methodology can be applied.

2.3 *Multicriteria Evaluation of Variants*

The MCDA methods can be divided into two major groups, such as [31]: multiple objective decision making methods and multiple attribute decision making methods. The first group is characterized by a huge number of variants, which is not given at the beginning of the decision process. They are designed and evaluated on the basis of a mathematical framework. As a result a level of multiple objective satisfactions is measured. In the second group of MCDA methods the number of variants is limited and it is known at the beginning of the decision making process. The final effect can be presented as a choice of the variant, e.g. Genetic Algorithm [32], sorting variants, e.g. UTADIS [33] or ranking variants, e.g. Electre III [12]. The presented methods use deterministic values of evaluation criteria and DM's preferences, as well.

The uncertain character of many problems related to the transportation network redesign, including non-deterministic performance information and the DM's preferences with a hesitation level to some extent, can also be encompassed by the decision aiding methods. The examples of those related to the ranking problematic are as follows: SMAA-III [34], stochastic PROMETHEE/GIS method [35], Fuzzy UTASTAR [36].

All the MCDA methods have many advantages, such as: precise modeling of decision maker's preferences, e.g. Electre III method [12], direct or indirect modeling of variants' performances, e.g. PROMETHEE method [37] and AHP method [13] respectively, possibility to solve the problems with a large number of variants, e.g. UTA method [38], possibility to model the stochastic criteria values, e.g. SMAA-III method [34], possibility to model DM's uncertain preferences, e.g. stochastic PROMETHEE/GIS method [35]. However, their application should be preceded by a thorough analysis of their suitability to the considered problem, the available data, DM's preferences as well as the expected result, e.g. ranking or sorting variants.

3 Proposed Methodology

The authors of the chapter propose a methodology that allows designing, modeling, simulating, evaluating and ranking the improvements of the existing transportation network. The methodology is composed of two phases, i.e. (1) simulation of the current state of the transportation network and its redesign scenarios, and (2) stochastic multiple criteria evaluation. Its scheme is presented in Fig. 1.

Each phase consists of four steps, which are as follows:

- Phase 1—simulation:
  - Step 1.1.—analysis of the transportation network.
  - Step 1.2.—macroscopic simulation model design and experiments.
  - Step 1.3.—design of traffic control rules.
  - Step 1.4.—microscopic simulation model design and experiments, results' analysis.
- Phase 2—stochastic multiple criteria decision analysis (MCDA):
  - Step 2.1.—multicriteria evaluation of redesign scenarios.
  - Step 2.2.—selection of MCDA method and computational experiments.
  - Step 2.3.—classification of rankings and stochastic ranking construction.
  - Step 2.4.—assessment of final solution and sensitivity analysis.

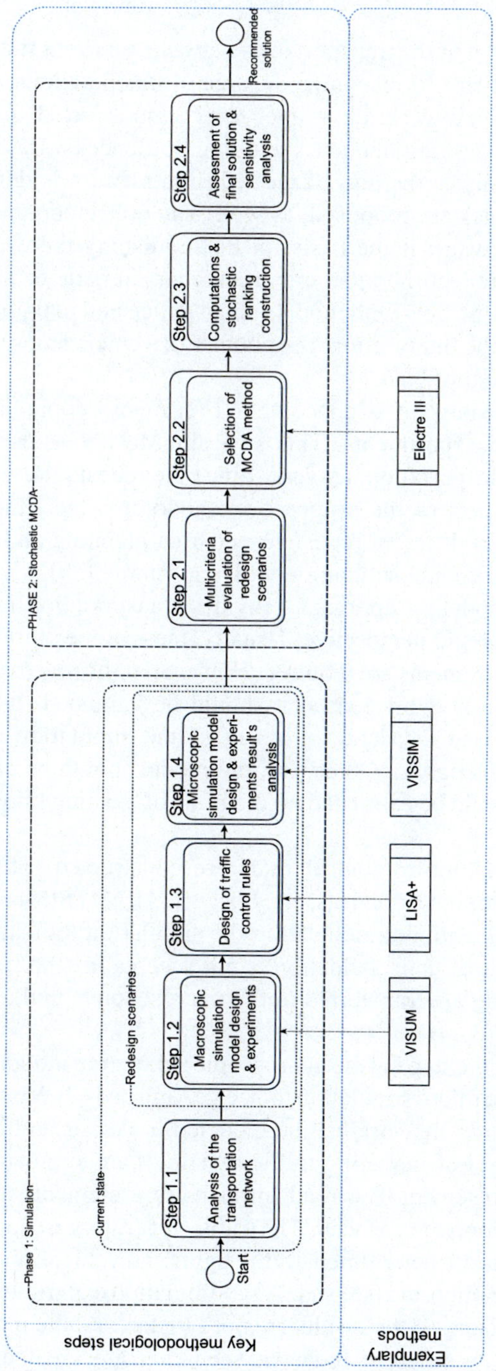

Fig. 1 Key phases and steps of the transportation network redesign methodology

3.1 Phase 1—Simulation

Modeling and simulation of the transportation network's current state begins with its analysis—step 1.1. It helps to recognize the area of its operations, the most important components of the network, the sequence of actions, etc. During this step the drawbacks of this network are defined, including bottlenecks. Based on the results of this preliminary analysis the aim of the redesign process is defined and potential directions of changes are proposed, as well. The collected general information about the analyzed network is the basis for distinguishing the evaluation criteria. These criteria should reflect the most important characteristic of the transportation network, including: traffic flow with modal split, average and maximum queue length of vehicles waiting at the intersection, the number of vehicle stops, total time losses resulting from vehicle stops.

Then, the step 1.2 starts. It is concentrated on macroscopic simulation model design and simulation experiments. This step aims to define the traffic intensity for junctions of the transportation network. Since the redesign of the transportation network is the key aspect of the proposed methodology, the simulation model is going to be used within different redesign scenarios planning and simulating. It is supported by the macroscopic simulation tool, e.g. Emme [39], Visum [26].

This traffic flow modeling approach provides transportation network planning with the wide and strategic perspective. Usually, the data need to be collected and the traffic-based measurements are required. However, if the macroscopic simulation model of the analyzed network exists, it should be validated. It is expected that the difference of the real data and the result of the simulation experiment is in the assumed range of variation. Otherwise, the model should be calibrated, i.e. the model parameters' should be estimated on the basis of existing knowledge or results of experiments [40].

During the step 1.3 traffic control rules are constructed. This step could be completed by the traffic signal controls, e.g. Lisa + [24], MITSIMlab [41]. The traffic signal operations are identified, modeled in the simulation tool and traffic controls are tested. If the result of their validation is not acceptable, the traffic controls are calibrated. Finally, they are transferred to the microscopic simulation tool and step 1.4 starts.

At the beginning of the step 1.4 the microscopic simulation model is designed. This step is supported by the micro simulation tool, e.g. Aimsun [42], Vissim [26]. Selected part of the transportation network can be modeled in details and its operations are simulated. On this level of modeling traffic dynamics are simulated and the single vehicle units are represented. The road infrastructure elements are identified and modeled in the simulation tool, as well. The traffic generators are introduced, priority rules at collision points are determined, traffic lights and vehicle detectors are located and traffic controls are imported (see step 1.3). Simulation experiments are carried out and the model is validated. If the results are not satisfactory, the model is calibrated.

During the phase 1 different redesign scenarios are created, including their assumptions and characteristics. The organizational and infrastructural changes of the transportation network's selected part are introduced. Based on that, the macrosimulation models are constructed (step 1.2). The modeled changes lead to the minimization of drawbacks in the current state of the transportation network. After modeling, the simulation experiments are carried out.

Then, new traffic control rules within the step 1.3 for the redesign scenarios are constructed. For each intersection, the set of operations is carried out, including: calculation of intergreen times between each signal groups, green intervals, intelligent logic control design, and detection design. As a result of this step, new traffic controls are exported to the microscopic simulation tool.

In the step 1.4 of phase 1, new microscopic simulation models are designed for each redesign scenario. The entire operations of the simulation models' construction, specified for the step 1.4 of the current state, are carried out for the redesign scenarios. Then, the simulation experiments are carried out.

Finally, the obtained results are analyzed. They are compared with the current state (step 1.4) and the aim of the research (step 1.1). If the level of improvement is not satisfactory, the procedure of the transportation network redesign should be repeated and the verification of the redesign's aim is recommended. At the end of this phase the results of the computations are accepted.

3.2 Phase 2—Stochastic MCDA

Based on the analysis of the simulation experiments' results, carried out in the phase 1, the current state of the transportation network and its redesign scenarios are evaluated according to the set of criteria defined in the step 1.1. This can be extended, according to the specific characteristics of the redesign scenarios, such as a new mode of transport introduced to the variant. The evaluation criteria are generated to compare the results of the simulation experiments.

The complex character of the network, its dynamic character changing the operations over time, and its representation in the simulation model, influence the final values of the criteria. The evaluations could be deterministic and stochastic. Moreover, the number of redesign scenarios, which is usually higher than two, and the number of criteria exceeding two, as well, make the final selection of the compromise variant difficult. In order to recommend the best scenario, the application of the MCDA method is proposed (step 2.2.). It is expected that the results of its application show the final ranking of variants and the solution of the decision problem, i.e. the most desirable improvement of the transportation network. However, based on the literature review and authors' experience, the selection of the MCDA method is usually made by the analysts' own preferences. This approach reduces among stakeholders a confidence in the calculation process and obtained results. Thus, there is proposed the selection of the MCDA method, which suits best the considered problem (Fig. 2).

The methodology is composed of four main steps, such as: (1) a comparative analysis of deterministic MCDA methods, based on the classification of these methods,

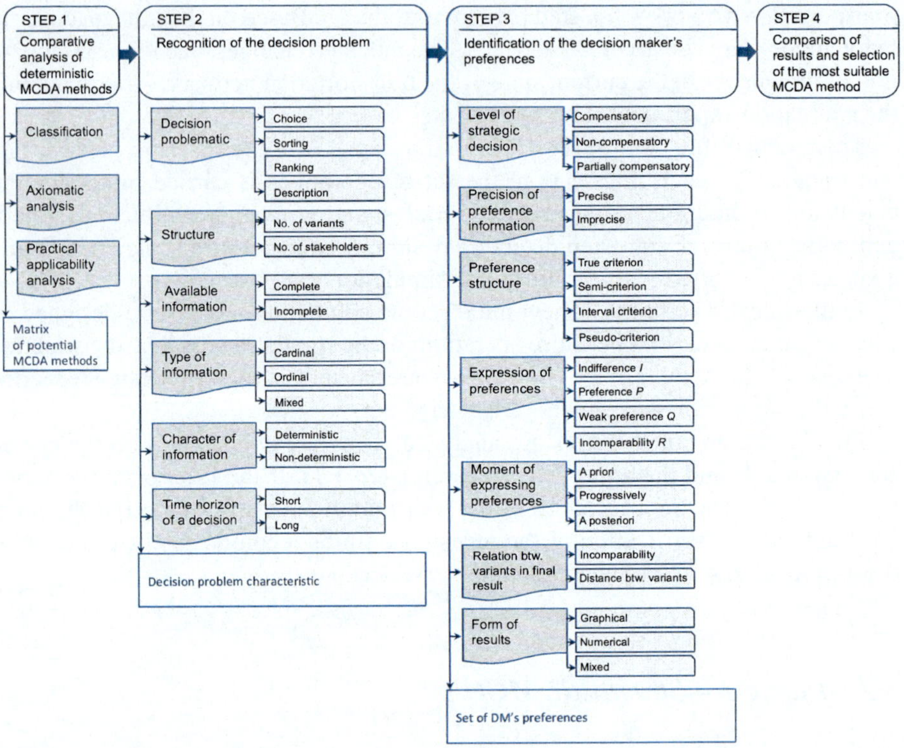

Fig. 2 The scheme of the most suitable MCDA method selection, based on [43]

their axiomatic analysis and practical applicability, resulting in the matrix of potential MCDA methods to be applied; (2) recognition of the decision problem, including its structure, decision problematic, available information and its type, as well as character, and time horizon of the decision; (3) identification of the DM's preferences, such as the level of strategic decision, precision of the preference structure, expression of preferences, moment of expressing them, expected relation between variants in a final result and the form of result; (4) comparison of results and selection of the most suitable MCDA method. A detailed analysis of the above mentioned aspects is presented by H. Sawicka [43]. As a result of step 2.2., the computations are carried out (step 2.3.).

The stochastic evaluations of variants are transformed into the deterministic values by random sampling (Fig. 3, level 1). These values, as well as the deterministic values of the other criteria, are input data for computations with an application of selected MCDA method (level 2). Finally, the set of rankings of variants is obtained. Since, there are some MCDA methods, with the preference structure presented as the quantitative information or qualitative (verbal) information, the application of probability theory (level 3a) or classification method (level 3b) is proposed. The result of the operations carried out on level 3 is the set of stochastic information, which is transformed into the stochastic ranking of variants. Its main characteristic

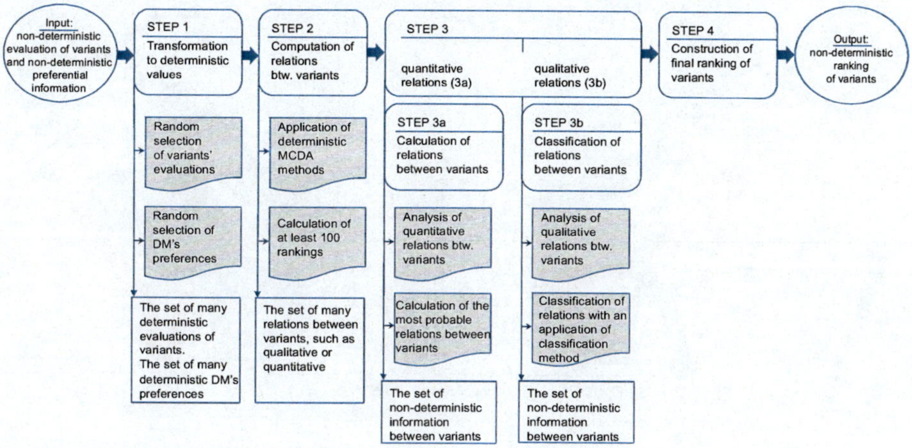

Fig. 3 Key steps of the stochastic MCDA methodology, based on [44]

is that the relation between variants, i.e. transportation network redesign scenarios, is presented as the probability of preference or indifference relation between them. More information about the application of this procedure is described by H. Sawicka [44].

The last step of the phase 2 of the proposed methodology is based on the evaluation of the final solution, represented by the stochastic ranking of variants. To strengthen the credibility of the aiding process, the sensitivity analysis is proposed. Finally, the best redesign scenario of the transportation network is recommended.

4 Application of the Proposed Methodology

The proposed methodology has been applied to solve the selected part of the transportation network problem in Poznan, Poland. The simulation phase starts with the analysis of the transportation network (step 1.1). Connected by a viaduct of 150 meters long, two channelized intersections are located along Kurlandzka Street. These intersepctions can be described as follows:

- Intersection 1—Kurlandzka St.—Wiatraczna St.—a four-leg intersection where the right of way is established for traffic flows moving between inlets i = 1 and i = 3;
- Intersection 2—Kurlandzka St.—Bobrzanska St.—a three-leg intersection where the right of way is established for traffic flows moving between inlets i = 5 and i = 6.

They are located in the southern part of Poznan in the area of one of the main communication routes of the city, i.e. Boleslawa Krzywoustego Street. This road connects the city's road infrastructure with the A2 motorway. For this reason, there

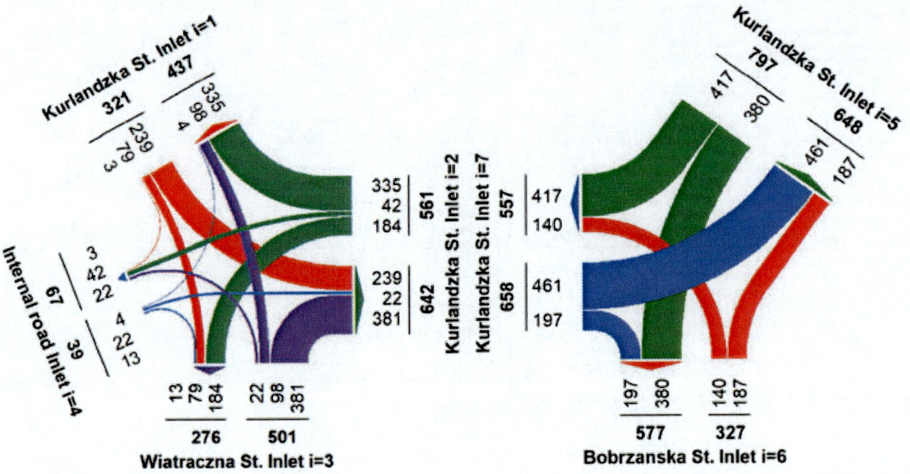

Fig. 4 One hour traffic flow diagram

is a high total traffic volume in the studied part of the network (Fig. 4), which is the cause of a constant traffic congestion and a low level of pedestrian safety.

Moreover, the existing infrastructure does not support cycling in any way. In order to improve the traffic conditions on the studied section of the network and to adapt it to the increasing needs of cyclists, 5 redesign scenarios denoted by variants have been analyzed. They are as follows:

- V1, which is the least investment variant—the existing organization has been expanded with new bicycle routes (Fig. 5a).
- V2 represents the extension of the variant V1 and includes coordinated traffic lights at both intersections. Signaling covers all flows of pedestrians and vehicles in the analyzed network (Fig. 5a).
- V3 assumes a complete change of the geometry of the intersection of Kurlandzka St. and Wiatraczna St. ($i = 1$) from the channelized intersection to the roundabout. A redesign of the intersection with Bobrzanska St. is the same as in V1 variant (Fig. 5b).
- V4 assumes a complete change of the geometry of the intersection of Kurlandzka St. and Bobrzanska St. ($i = 2$) from the channelized intersection to the roundabout. A redesign of the intersection with Wiatraczna St. is is the same as in V1 variant (Fig. 5c).
- V5 represents the most evolutionary scenario with a complete redesign of the geometry of both intersections, i.e. from the channelized intersections to the roundabouts (Fig. 5d).

Those variants are represented in a microsimulation model (step 1.2). In the matter of a work related to traffic modeling issues, the authors use PTV Vissim software, whereas in the part focused on designing the traffic light program (V2) Lisa + software is utilized (step 1.3).

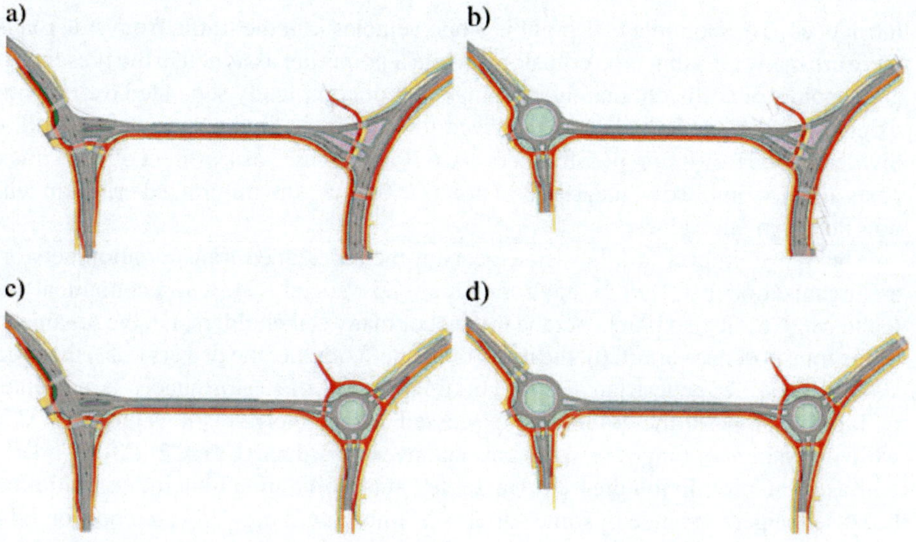

Fig. 5 Redesign scenarios of two intersections, **a** variant 1 (V1) and 2 (V2), **b** variant 3 (V3), **c** variant 4 (V4), **d** variant 5 (V5)

Each of the variants (V1-V5) is subjected to a series of twenty simulations (step 1.4). Each of them uses a different value of the Vissim's Random Seed parameter to account for the stochastic nature of vehicle movement on the road network.

Every single simulation experiment lasts 4500 s. Each experiment is divided into two phases: warm-up period and the main part of the experiment. During warm-up period lasting 900 s, the network is filled with vehicles to get stabilized traffic conditions. The main part of the experiment lasts 3600 s. All movements' data of vehicles generated in this time interval are evaluated and collected for a detailed analysis. Vehicles generated in the warm-up period are not taken into consideration while the simulation results' analysis.

Step 1.4 of the proposed methodology results in the data analysis. Then, in step 2.1 each variant is evaluated by the family of criteria C1*i*—C6. C1*i* is a queue length at the *i*-th road network inlet ($i = 1, 2, \ldots, 7$) [m], the minimized criterion with non-deterministic values. C2 is a delay per vehicle in the network [s], the minimized criterion with non-deterministic values. C3 is a fuel consumption per vehicle [l/100 km], the minimized criterion with non-deterministic values. C4 represents pedestrian and cyclist safety level, measured in points $< 0 \div 1 >$, the maximized criterion with deterministic values. Each inlet is evaluated in terms of pedestrian and cyclist safety using a scale from 0—no pedestrian or cyclist crossing to 4—short, signalised pedestrian and cyclist crossing across the one-way road. A total score achieved for all 7 inlets in the analysed network is divided by the maximum possible score. C5 is a safety of vehicle traffic flow, measured in points $< 0 \div 1 >$, the maximized criterion with deterministic values. Each inlet is evaluated in terms of collisions with the other traffic flows. For this purpose, a scale from 1 to 3 is introduced. The example of 1 point is when vehicles join the traffic from side roads, there are many crossing type collisions, while 3 points are assigned in the presence of a few points of collision, mainly merging type, or completely separated traffic flows using traffic lights. A total score achieved for all 7 inlets in the analysed network is divided by the maximum possible score. C6 is a combined assessment of investment costs and its complexity, measured in points $< 0 \div 1 >$, the minimized criterion with non-deterministic values.

This set of criteria fulfills such aspects of the redesigned transportation network as: organizational (C1i, C2), environmental (C3), social (C4, C5), economical and technical (C6). It also points out the interest of many stakeholders, i.e. the organizers of the transport network (C6), the users of the network, i.e. the drivers (C5), the other stakeholders, i.e. pedestrians (C4). The final matrix of performances is presented in Table 1. Some of the criteria are expressed as the deterministic values (C4, C5), while the others as ranges of variations and average values (C1*i*, C2, C3).

Based on information presented in Table 1 it is hard to point out the best solution, because the performance of some variants is ambiguous, e.g. V5 is the best on C14, but it is the worst on C6. The situation becomes more complex since many criteria are expressed as non-deterministic values.

Thus, in the next step (2.2) of the computational experiments the most suitable MCDA method has been selected. There were considered ranking methods, like AHP, Electre III [12], Promethee I [37], Promethee II [37], UTA [38], since the decision problem has been formulated as the hierarchy of variants with a recommendation of the best scenario. The comparative analysis of deterministic MCDA methods, matching between the methods and the decision problem, matching between the methods and DM's preferences resulted in the selection of Electre III method. The most important argument for this choice was the expression of preferences, including the veto threshold, as well as the possibility of the incomparability relation between variants in the final result.

In the step 2.3 the preferences of the decision maker, who is an analyst in the considered problem, have been collected. In the ELECTRE III method based on the outranking relation, the DM's preferences are determined by the indifference q_j, preference p_j, and veto v_j thresholds and weights w_j for each criterion j are defined (Table 2).

The most important criteria are C4 and C5 with the highest weights 10. These criteria are connected with safety of the stakeholders. The next in the hierarchy of weights is value 7 assigned to C2, which is a delay per vehicle in the network. The total weight of criteria C1*i* is 5. This value was divided by the total number of *i*, to show the importance of these criteria as all. The investments (C6) and fuel consumption (C3) are the criteria of the least importance for the DM. The thresholds also vary and the model of preferences includes the values for all of them, e.g. the DM recognizes variants with the difference of fuel consumption (C3) at most 0,5 l/100 km as indifferent, while the difference higher than 0,5 l/100 km but lower or equal to 1 l/100 km shows the weak importance of one variant (with lower fuel consumption) over the other. The relation between variants with the difference of fuel consumption higher than 1 l/100 km and lower than 5 l/100 km is of strong preference. The variants

Table 1 The final matrix of variants' performances

Criteria			Variants				
Name	Unit preference	Value	V1	V2	V3	V4	V5
C11	[m] min	average	142	71	60	50	43
		[min; max]	[23; 512]	[46; 117]	[26; 332]	[15; 325]	[26; 65]
C12	[m] min	average	80	107	99	51	52
		[min; max]	[20; 236]	[79; 184]	[34; 751]	[15; 351]	[26; 84]
C13	[m] min	average	13	93	267	11	73
		[min; max]	[6, 24]	[69; 152]	[65; 757]	[6, 16]	[38; 219]
C14	[m] min	average	8	20	4	7	1
		[min; max]	[0; 21]	[14, 26]	[0; 20]	[0; 22]	[0; 9]
C15	[m] min	average	14	174	18	257	231
		[min; max]	[0; 25]	[112; 294]	[0; 160]	[143; 405]	[134; 349]
C16	[m] min	average	22	55	35	69	60
		[min; max]	[12, 39]	[41; 87]	[11; 261]	[34; 119]	[38; 127]
C17	[m] min	average	300	66	277	196	164
		[min; max]	[197; 397]	[42; 98]	[181; 392]	[113; 270]	[94; 242]
C2	[s] min	average	42	39	50	31	29
		[min; max]	[21; 97]	[37, 42]	[21; 134]	[23; 55]	[24, 36]
C3	[l/100 km] min	average	11	11	12	11	11
		[min; max]	[10, 14]	[10, 11]	[10, 16]	[10, 12]	[10, 11]
C4	[-] max	–	0.29	0.63	0.39	0.43	0.54
C5	[-] max	–	0.60	0.93	0.86	0.74	1.00
C6	[-] min	[min; max]	1	[2, 5]	2.5	2.5	4

Table 2 Model of the DM's preferences in Electre III method

w & t[a]	Criteria											
	C11	C12	C13	C14	C15	C16	C17	C2	C3	C4	C5	C6
w_j	0.7	0.7	0.7	0.7	0.7	0.7	0.7	7	3	10	10	4
q_j	15	10	15	10	15	15	10	5	0.5	0.08	0.08	0.5
p_j	40	40	40	30	40	40	40	15	1.0	0.27	0.21	2.5
v_j	200	100	200	100	200	200	100	30	5.0	0.40	0.50	4.0

[a] *w & t—weights and thresholds*

with the difference of the fuel consumption at least 5 l/100 km are recognized as incomparable.

Since the character of information, presented in the matrix of performances (Table 1), is non-deterministic the computations have been carried out with the approach presented in Fig. 3. The non-deterministic evaluations of variants have been transformed into the deterministic values by random selection. Based on them the series of 150 calculations have been carried out with the application of Electre III method. As a result 150 rankings of variants has been generated. Due to the nature of the selected Electre III method, the relations between variants are qualitative, i.e. indifference, preference and incomparability. They have been classified to six decision classes, including *I*—indifference relation between variants, *P*—strong preference relation between variants, *P*—reciprocal of strong preference relation between variants, *R*—incomparability relation between variants, *Q*—weak preference relation between variants and *Q*—reciprocal of weak preference relation between variants. The result is presented in Fig. 6.

This is the hierarchy of variants with the leader on the top and the worst variant at the bottom (Fig. 6a) with the probability of preference *P* relations. The precise information of probabilities of occurrence of particular relations between variants is also demonstrated (Fig. 6b). For example, the final relation between variants V2 and

a)

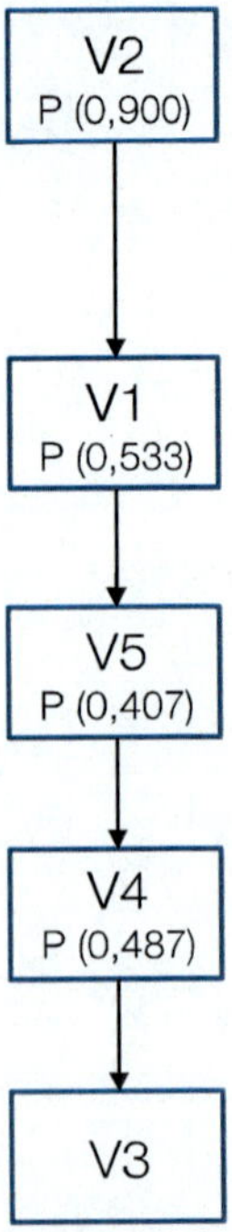

b)

	V1	V2	V3	V4	V5
V1	I (1,000)	P-(0,900) I (0,060) P (0,040)	P (0,760) I (0,147) P-(0,093)	P (0,740) R (0,100) P-(0,087) I (0,073)	P (0,533) I (0,240) P-(0,167) R (0,060)
V2	P (0,900) I (0,060) P-(0,040)	I (1,000)	P (0,907) P- (0,073) I (0,013) R (0,007)	P (0,893) I (0,073) R (0,033) P- (0,027)	P (0,907) P- (0,053) I (0,033) R (0,007)
V3	P-(0,827) I (0,147) P (0,027)	P- (0,907) P (0,073) I (0,013) R (0,007)	I (1,000)	P- (0,487) P (0,267) I (0,227) R (0,020)	P- (0,620) I (0,187) P (0,180) R (0,013)
V4	P-(0,740) R (0,100) P (0,087) I (0,073)	P- (0,893) I (0,073) R (0,033) P (0,027)	P (0,487) P- (0,267) I (0,227) R (0,020)	I (1,000)	P- (0,407) P (0,300) I (0,187) R (0,107)
V5	P-(0,533) I (0,240) P (0,167) R (0,060)	P- (0,907) P (0,053) I (0,033) R (0,007)	P (0,620) I (0,187) P- (0,180) R (0,013)	P (0,407) P- (0,300) I (0,187) R (0,107)	I (1,000)

Fig. 6 The results of the step 2.3, **a** a stochastic ranking of variants, **b** a stochastic ranking matrix representing the mutual relations between variants

V1 is *P* (preference) and the probability of its occurrence equals 0,900. The relations of *I* (indifference) and *P*- (reciprocal of preference) is also probable, but with the low probability level, i.e. 0,060 and 0,040, respectively.

Next, the final ranking has been analyzed (step 2.4). The leader is variant V2, which represents slight changes in the existing transportation network, improving safety of pedestrians, cyclists and vehicle drivers, as well as changing the organizational rules with coordinated traffic lights. It is worth mentioning that this solution is placed with a high distance from the remaining redesign scenarios. It is indicated by a high probability of the preference relation, i.e. with a high degree of certainty. The second position in the ranking has the variant V1 with expansion of new bicycle routes. Close to this solution in the final hierarchy is V5 representing a complete redesign of the geometry of both intersections. The fourth postion in the ranking has V4 and the fifth—V3. These two last variants assume the complete change of the geometry of the selected intersections.

Finally, the sensitivity analysis has been carried out. There have been analyzed final rankings for minimum, i.e. optimistic, maximum, i.e. pessimistic and average values of criteria evaluating variants (Fig. 7).

The results of the sensitivity analysis show that two rankings of variants are similar. However, the hierarchy of redesign scenarios for the optimistic data (Fig. 7a) is different. The leader is V5 representing the most radical changes. It means that this redesign scenario would be the best option for the best situation in the transportation network. Variant V5 outranks variant V3 and variant V4 is on the third position. The worst redesign scenarios are V1 and V2 assuming the evolutionary changes in the transportation network. The pessimistic conditions (Fig. 7b), as well as the average

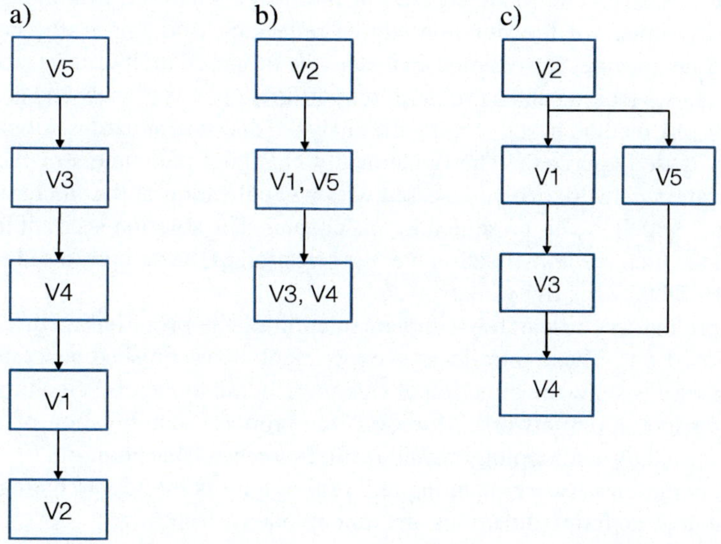

Fig. 7 The results of the step 2.4—ranking of variants for the minimum (**a**), maximum (**b**) and average (**c**) values of ranges of variations calculated with an application of Electre III method

values of criteria (Fig. 7c) result in the V2 as the best scenario and variant V5 is on the second position with V1 as, respectively, indifferent and incomparable solution. The worst variants are V3 and V4, which are indifferent regarding the maximum values of criteria. V3 is incomparable with V5, and preferred to V4 in the ranking based on the average values of evaluation criteria.

Based on the analysis of the final stochastic ranking of variants and the sensitivity analysis the authors of this chapter recommend variant V2 as the redesign scenario of the analyzed part of the transportation network. It is a leader with high confidence level, which also performs well in difficult network conditions.

5 Conclusions

In this chapter the methodology of designing, modeling and verifying changes of the existing transportation network has been presented. The aim of this approach is to improve the actual transportation network. The decision problems considered within this research area are characterized by the complexity. Thus, the presented methodology is composed of two phases, i.e. connected with simulation and stochastic MCDA. These phases are divided into eight steps linking three modeling approaches, such as: the macroscopic, microscopic and traffic signal controls, as well as multiple criteria decision aiding. The proposed methodology can be implemented to find the strategic solution (macroscopic part of the approach). It also provides the general overview of the problem. Depending on the considered situation, a selected part of the transportation network can be analyzed. The microscopic approach would facilitate this step. The modeled area of the network can be examined in details and the dynamism of the situation can be reflected. Modeling traffic controls for intersections specifies the detailed analysis. The proposed multicriteria evaluation of variants allows looking at the problem from different points of view and the selection of the MCDA method best matching the analyzed decision situation is the advantage of the proposed approach. The dynamically changing parameters of the modeled transportation network can be assessed with an application of the stochastic MCDA procedure, which is also presented in this chapter. The solution is a ranking of variants and the final recommendation, i.e. the selection of the compromised solution, is supported by the sensitivity analysis.

The preliminary research on implementation of the proposed methodology has been carried out. These introductory experiments have finished successfully. The obtained results show the direction of changes leading to the expected improvement of the transportation network. Moreover, the proposed combination of simulation tools and MCDA methodology would result in a reasonable solution.

Transportation network planning and redesigning is the research area, which is still evolving. Its future directions are, among others, as follows:

- verification of the methodology presented in this chapter on transportation networks with different size (arcs and nodes) and characteristics;

- robustness analysis, which is the subject of research of X. Sun, V. Gollnick and S. Wendelt [45];
- evaluation of the proposed transportation network redesign scenarios, including different points of view of stakeholders, i.e. different models of decision making preferences.

References

1. Law AM, Kelton WD (2000) Simulation. Modeling and analysis. McGraw Hill, New York
2. Fishman GS (2001) Discrete-Event simulation. Modeling programming, and analysis. Springer, New York
3. Macioszek E, Sierpiński G (2020) Charging Stations for electric vehicles—Current situation in Poland. In: Mikulski J (ed) Research and the future of telematics, communications in computer and information science, vol 1289. Springer, Heidelberg, pp 124–137
4. Macioszek E, Sierpiński G, Staniek M (2017) Analysis of trends in development of freight transport logistics using the example of Silesian Province (Poland)—a case study. Transp Res Proc 27:388–395
5. Földes D, Csiszár C (2018) Framework for planning the mobility service based on autonomous vehicles. In: Smart cities symposium, 24–25 May 2018. Prague, Czech Republic, pp 15–20. https://doi.org/10.1109/SCSP.2018.8402651
6. Friesz TL, Cho H-J, Mehta NJ, Tobin RL, Anandalingam G (1992) A simulated annealing approach to the network design problem with variational inequality constraints. Transp Sci 26(1):18–26
7. Leblanc JL (1975) An algorithm for the discrete network design problem. Transp Sci 9(3):183–199
8. Mizuno H, Okamoto T, Koakutsu S, Hirata H (2014) A method for design of a growing complex network. Electron Commun Japan 97(1):70–81
9. Abid NM, Hussain SS (2017) Transportation network planning using simulation. In: Proceedings of the 2nd IEEE International conference on intelligent transportation engineering, Singapore, September 1–3, pp 272–279
10. Abdelghany K, Mahmassani H (2001) Dynamic trip assignment-simulation model for intermodal transportation networks. J Transp Res Board 1771:52–60
11. Bloomberg L, Dale J (2000) Comparison of VISSIM and CORSIM traffic simulation models on a congested network. J Transp Res Board 1727:52–60
12. Roy B (1985) Methodologie Multicitere d' Aide a la Decision. Economica, Paris (in French)
13. Saaty TL (1980) The analytic hierarchy process: planning, priority setting, resource allocation. McGraw-Hill, New York
14. Vincke P (1992) Multicriteria decision-aid. Wiley, Chichester
15. Kicinski M, Bieńczak M, Fierek S, Merkisz-Guranowska A (2019) A multi-criteria decision making approach for the evaluation of roads and streets system in Gniezno. In: ICCPT 2019: Current problems of transport: proceedings of the 1st international scientific conference, May 28–29, pp. 22–31. Scientific Publishing House "SciView", Ternopil, Ukraine
16. Belton V, Stewart TJ (2003) Multiple criteria decision analysis. An integrated approach. Kluwer Academic Publishers, Dordrecht
17. Mitchell T (1997) Machine learning. McGraw-Hill, New York
18. Bevan TA, Mason R, McKenzie JA (2008) Context sensitive design challenges for major suburban arterial street projects. In: 2nd Urban Street Symposium: uptown, downtown, or small town: designing urban streets that work, July 28–30, 2003, Anaheim, California

19. Cantisani G, Loprencipe G, Primieri F (2011) The integrated design of urban road intersections: a case-study. Conference Paper, ICSDC 2011: integrating sustainibility practices in the construction industry
20. Gaca S, Suchorzewski W, Tracz M (2014) Inżynieria ruchu drogowego. Teoria i praktyka, Wydawnictwo Komunikacji i Łączności, Warszawa (in Polish)
21. https://ec.europa.eu/transport/road_safety/specialist/knowledge/road/getting_initial_safety_design_principles_right/junctions_en
22. Wiedemann R (2019) Evaluation of the Improvement of Organisation and Traffic Control System on Selected Part of the Transport Network. Współczesne problemy transportu: opracowanie monograficzne. T. 5, Sterowanie ruchem. Koło Naukowe Systemów Komunikacyjnych Politechniki Krakowskiej im. T. Kościuszki, pp 20–33 (in Polish)
23. Demir HG, Demir YK (2020) A Comparison of traffic flow performance of roundabouts and signalized intersections: a case study in Nigde. The Open Transp J 14(1):120–132
24. www.schlothauer.de
25. Lisa+ver. 5.2 User Manual
26. www.ptvgroup.com
27. www.stadtraum.pl
28. Hoogendoorn SP, Bovy PHL (2001) State-of-the-art of vehicular traffic flow modelling. J Syst Control Eng 215(4):283–303
29. Mohan R, Ramadurai G (2013) State-of-the-art of macroscopic traffic flow modelling. Int J Adv Eng Sci Appl Math 5(2–3):158–176
30. Wiedemann R (1974) Simulation des Straßenverkehrsflusses. Schriftenreihe des Instituts für Verkehrswesen der Universität Karlsruhe (in German)
31. Kahraman C (ed) (2008) Fuzzy multi-criteria decision-making. Theory and applications with recent developments. Springer, New York
32. Goldberg D (1989) Genetic algorithms in search. Optimization and machine learning. Addison-Wesley, Boston
33. Manshadi ED, Mehregan MR, Safari H (2015) Supplier classification using UTADIS method based on performance criteria. Int J Acad Res Bus Social Sci 5(2):31–45
34. Tervonen T, Figueira J (2008) A survey on stochastic multicriteria acceptability analysis methods. J Multi-Criteria Decis Anal 15(1–2):1–14
35. Marinoni O (2005) A stochastic spatial decision support system based on PROMETHEE. Int J Geogr Inf Sci 19(1):51–68
36. Patiniotakis I, Apostolou D, Mentzas G (2011) Fuzzy UTASTAR: a method for discovering utility functions from fuzzy data. Expert Syst Appl 38(12):15463–15474
37. Brans JP, Vincke P, Mareschal B (1986) How to select and how to rank projects: the PROMETHEE method. Eur J Oper Res 24(2):228–238
38. Jacquet-Lagrèze E, Siskos Y (1982) Assessing a set of additive utility functions for multicriteria decision making: the UTA method. Eur J Oper Res 10(2):151–164
39. www.inrosoftware.com
40. Edward JD (1992) Transportation planning handbook. Pearson College Division, London
41. https://its.mit.edu/software/mitsimlab
42. www.aimsun.com
43. Sawicka H (2020) The methodology of solving stochastic multiple criteria ranking problems applied in transportation. Transp Res Proc 47:219–226
44. Sawicka H (2017) Artificial intelligence in stochastic multiple criteria decision making. In: Trivedi SK, Dey S, Kumar A, Panda TK (2017) Handbook of research on advanced data mining techniques and applications for business intelligence. IGI Global, Chapter 19, pp 315–340
45. Sun X, Gollnick V, Wendelt S (2017) Robustness analysis metrics for worldwide airport network: a comprehensive study. Chin J Aeronaut 30(2):500–512

Evaluating the Efficiency of Regional Transport Network

Firas Alrawi and Faisal A. Mohammed

1 Introduction

The regional transportation policies have an essential and long-term impact on the quality of life and economic welfare [1]. The regional transportation networks do not only provide for the movement of passengers and goods; it also forms the patterns for development and economic activities through accessibility and linking the settlements [2]. Without connecting settlements with each other, the regions cannot develop. Human settlements (urban and rural) are linked to each other by multiple and varied activities. These activities consist of multiple types of movement and transportation, and the result of this interaction is a stable system in place [3].

1.1 Regional Transport Network

A region is a place where a group of elements is organized into relationships, as Steiner described it. These elements are dividing into natural (topography, water, natural vegetation, resources, etc.) and environmental (population and their densities, lifestyle, social structure, daily activities, workforce, and market) [3]. The relationships in a region cannot continue well without efficient regional road networks. Ndulu believes that the geographical fragmentation of settlements has significant disadvantages for development in Africa's regions. He called for strengthening these regions'

F. Alrawi (✉) · F. A. Mohammed
Urban and Regional Planning Centre for Postgraduate Studies, University of Baghdad, Baghdad, Iraq
e-mail: dr.firas@uobaghdad.edu.iq

F. A. Mohammed
e-mail: faissal.abd1200b@iurp.uobaghdad.edu.iq

G. Sierpiński and E. Macioszek (eds.), *Decision Support Methods in Modern Transportation Systems and Networks*, Lecture Notes in Networks and Systems 208,
https://doi.org/10.1007/978-3-030-71771-1_2

infrastructure to promote it by establishing modern transport networks that link them within the same region or neighboring regions and neighboring African countries. He indicated that regional transportation networks provide access to economic inputs of knowledge, resources, and technology and reduce barriers to the mobility of goods and people [4].

Monitoring the performance of regional transport networks and their continuous maintenance and making improvements and development through establishing a multimodal transport network contributes to reducing transmission delays, increases reliability, maintains safety and provides transportation options with lower costs [5].

The establishment and planning of regional roads for development are not only by preparing highway and developing property services and investments; it requires advancing procedures for financing, operating, and managing the transportation network and all utilities of the region [4].

In other words, when constructing new regional roads, a policy should be followed that contributes to preserving the energy, environmental and aesthetic quality of the project, strengthening the economy, enhancing social justice, and making societies more liveable [2]. Regional road network performance influences general policy matters like smart growth, air quality enhancement, resource consumption, justness, development, security, and so on [6].

Therefore, transportation on both the infrastructure and services sides is the main factor in the region's sustainable development, and the effects of development are usually direct and indirect. One of the foundations of sustainable development is establishing regional transport networks that contribute to obtaining economic, social, and environmental goals [7–11].

1.2 Regional Transportation Network and Development

One of the regional development policy's essential goals is to develop the region by raising its contribution to the national product. This policy is based on reorganizing the regions by distributing infrastructure and services towards all the regions' settlements. The imbalance in the sizes and distribution of settlements in the same region or between regions leads to the emergence of attractions and other areas for population expulsion (i.e., developed and undeveloped areas) [3].

The development of settlements can be measuring through activities concentration, and trips are a function of activity by determining their total over a certain period in a particular region and through regional routes [12].

Therefore, the volume of the region's traffic flows and the nature of the regional transport network determine the scope of its impact, which has a fundamental role in the regional interaction between various cities and settlements [2, 13, 14]. This interaction shows the importance of regional transport in revitalizing and activating the regional link between cities by investing the available resources at the regional level [15].

The regional system and the movement of individuals within it are caused by the fact that individuals are not self-sufficient, which calls for cooperation, exchange, and interaction with others inside and outside the region. If each individual produced what he consumed, then the need to live inside the settlements would not be [3]. Individuals and companies make their transportation decisions through a wide range of options. These decisions are usually influencing by the built environment and the availability of infrastructure [13].

An efficient regional transportation system represents the backbone for any region [16] because regional economic growth and development depend on market access for producers and customers, which can be achieved with a strong infrastructure for the regional transportation system [17].

The sufficient flexibility in the regional transportation system encourages customers and producers to use it through the possibility of changing the road and/or means of transportation, and this can only be achieved with adequate alternatives in terms of modern transport networks and various means of transport [18].

It is known that transportation has positive effects in general, but there are other effects that transportation has concerning the development process. The literature review indicates three possible relationships with the presence of transportation: (1) a positive impact through the expansion of activities due to the availability of road networks; (2) a permissive impact as the road networks do not contribute directly and independently to the creation of economic growth; (3) a negative impact when excessive investment in transportation networks reduces the possibility of economic activity and thus reduces the absolute incomes of individuals [19].

However, as mentioned above, transportation has visible and unforeseen benefits. The Economic Development Research Group study in Portland, Oricon, USA, indicated the effects of crowding in the city's development through the ability to maintain economic growth, and came in its conclusions:

Failure to improve the state of transportation networks in the state may lead to losses estimated at $ 844 million per year by 2025, representing $ 782 per family; this equates to 28 travel hours for each family annually. Also, it is expected that every dollar spent in the transportation sector can bring in two dollars of benefit [7].

2 Materials and Methods

2.1 Dujail Region

Dujail city is located in the north of Baghdad Governorate, 50 km away, as it represents the southern part of Salah al-Din Governorate (SG) and one of eight main cities on it. Dujail has several main roads that penetrate it, and the city is considered a linked node between the governorate of Baghdad in the south and SG on the one hand, and the northern governorates on the other hand. The city has taken a linear shape, following the main roads that cross it. It suffers from many traffic problems resulting from local and through-traffic, including environmental and social problems related to traffic accidents, traffic congestions, and various pollutants.

2.2 *Data Collection and Analysis*

To test the relationship between Dujail city and its region and to know the city's regional transport network's efficiency, a questionnaire form was formulated for the city's residents to identify the nature of the relationship between the city and its region. The number of electronic questionnaire forms obtained reached more than a thousand. The orientation through the questionnaire was to cover a vast and diverse segment of the city's population. The sample taken gave a broad view of the extent of the city's relationship with its regional surroundings, including villages, cities, and governorates.

3 Results and Discussion

This part of the research divides into two aspects; the first one dealt with a set of general indicators for the regional transportation network of Dujail, while the second one discussed the connection of the city of Dujail with its regional surroundings.

3.1 *General Indicators of Dujail Regional Transportation Network*

These paragraphs deal with the nature of regional trips in the region and the state of the transport infrastructure, as seen by road users and traffic accidents in the region.

Trips Analysis. The daily, weekly, monthly, and yearly trips were analyzed to identify the regional connectivity of Dujail city and the efficiency of the surrounding regional transportation network. Table 1 shows the regional travel patterns for Dujail residents, according to its frequency and purpose. It is noticed that the percentage of daily trips for the city's population is 19%, which is predominantly for work trips. The percentage of trips that were more than once a week was 23%. The weekly trips recorded 18%, the monthly 16%, and the annual 24%, Fig. 1. Most of the trips characterized as trips for work by 41%, followed by trips for religious by 17%, educational trips by 14%, shopping trips by 8%, then trips for health and other purposes by 7% each, finally, trips for social and recreational purposes at a percentage of 6%, as shown in Fig. 2.

Table 1 Distribution of frequency regional trips for Dujail residence

	Number of trips						
Trip purpose	Daily	More than one in a week	Weekly	Monthly	Yearly	Total trips	%
Work	129	136	86	56	46	453	41
Shopping	14	17	21	23	18	93	8
Education	48	36	27	31	17	159	14
Social & recreation	4	13	8	28	12	65	6
Health	3	9	21	14	26	73	7
Religious	5	8	34	19	121	187	17
Other	9	21	8	11	31	80	7
Total trips	212	240	205	182	271	1110	100
%	19	23	18	16	24	100	

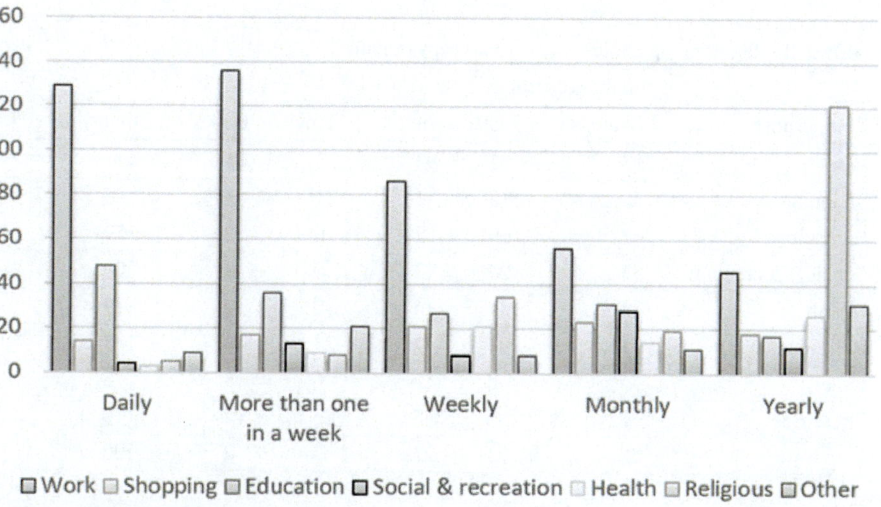

Fig. 1 Distribution of frequency regional trips for Dujail residence

The previous table indicates that since Dujail is a city that grew out of villages dependent on an agricultural basis, it led to a lack of job opportunities and its dependence on its regional surroundings. Also, its residents who are made regular trips for religious purposes, whether weekly or annual, toward some cities that include religious symbols such as Baghdad, Samarra, and Balad, which increased the percentages of trips for this purpose.

About transport modes, Table 2 and Fig. 3 shows that most of Dujail city residents depend on private cars for their regional trips, at a percentage of 46%.

The other transport means were varying between 23% for taxis, 22% for minibus, 7% for trucks, and 2% for high occupancy vehicles (HOV). The reason for the use of

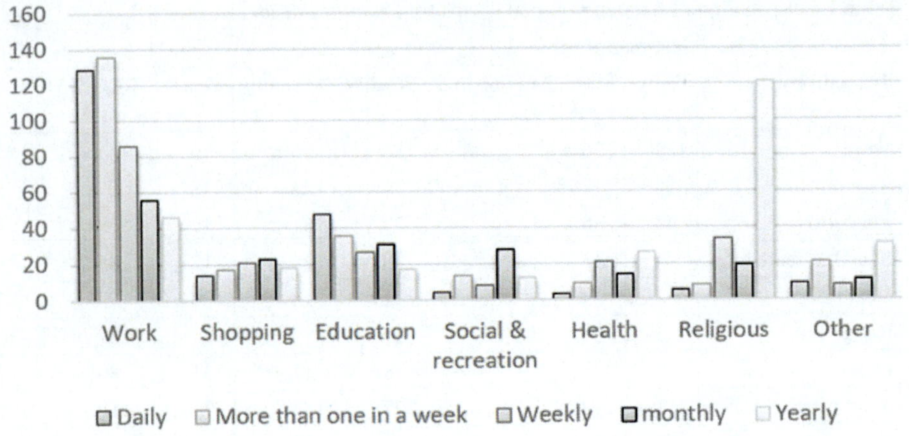

Fig. 2 Regional trips purpose for Dujail city

Table 2 Different mode choice for regional trips in Dujail

	Number of trips						
Trip purpose	Private car	Taxi	Minibus	Truck	H.O.V.	Total trips	%
Work	213	101	93	33	11	451	41
Shopping	38	13	11	21	3	86	8
Education	68	39	52	0	0	159	14
Social & recreation	33	24	13	0	0	70	6
Health	28	37	14	0	0	79	7
Religious	91	31	42	23	0	187	17
Other	44	8	17	6	3	78	7
Total trips	515	253	242	83	17	1110	100
%	46	23	22	7	2	100	

private cars a lot by the residents of Dujail is that the area suffers from a shortage of public transport means, as there are few minibusses and taxis. Therefore, the residents prefer the private cars because it is more comfortable, flexible, and cheaper at times, and this is shown in Table 3 and Fig. 4, which illustrate the reasons for choosing the regional transport means for the city's residents.

As evident from the previous table, the population's preference directed towards comfort and a sense of safety by 46%. The means from door to door (shorter travel time) was 24%. In addition to the moderate transportation price compared to the rest of the available public transport means. In the current transportation conditions, the most effective means to achieve these advantages is the private vehicle, which has occupied 74% of the population's preference.

As for the obstacles that city residents face in their regional trips, Table 4 and Fig. 5 illustrate them.

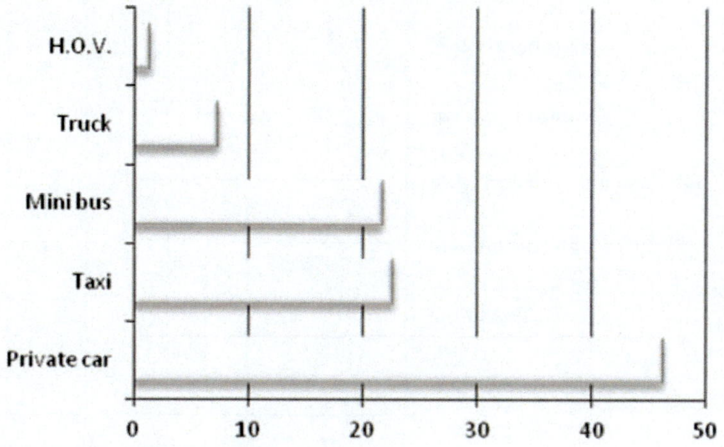

Fig. 3 Different mode choice for regional trips in Dujail

Table 3 Reasons for preferring different means of transportation for Dujail residents during regional trips

	Number of trips							
Trip purpose	Cheaper	Short time	Low income	Safer and more comfortable	Has not owned a car	No other choices	Total trips	%
Private car	143	166	28	508	18	21	884	67
Taxi	34	126	0	60	7	90	317	24
Minibus	23	3	33	2	17	8	86	6
Truck	7	2	0	11	0	3	23	2
HOV	0	0	0	0	0	10	10	1
Total trips	207	297	61	581	42	132	1320	100
%	16	22	5	44	3	10	100	

As it turns out that the most significant obstacles were the severity of traffic congestions by 34%, as for poor road conditions, it was 27% due to the old road network system for this region and the lack of maintenance for it for a long time. Other reasons distribute as follows: lack of parking spaces by 11%, absence of public transportation by 26%, and unavailability of a private car by 2%.

Transport infrastructure assessment. For evaluation, the infrastructure of the entire transportation network system for the Dujail city region, a group of questions, directed to the study sample related to construction road status, traffic signs, and road furniture. Table 5 shows the results of Dujail residents'observations about the regional roads linkage with their city.

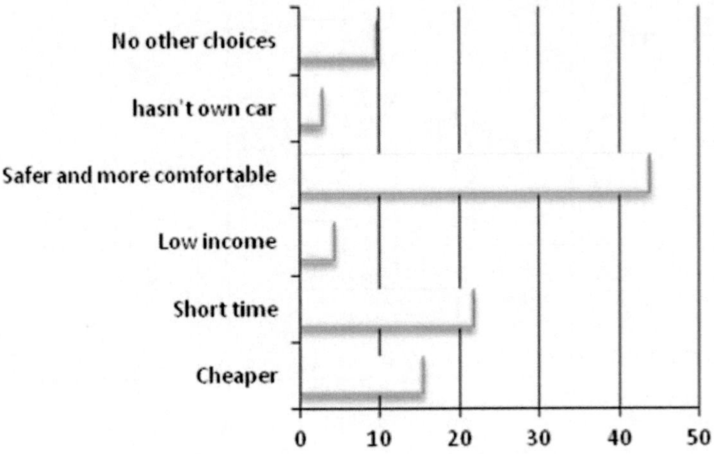

Fig. 4 Reasons for preferring different means of transportation for Dujail residents during regional trips

Table 4 The obstacles down to the final destination

Obstacles	%
Traffic congestion and traffic obstacles	34
The lack of public transportation means	26
The lack of cars parking	11
Road conditions	27
Not owning a car	2

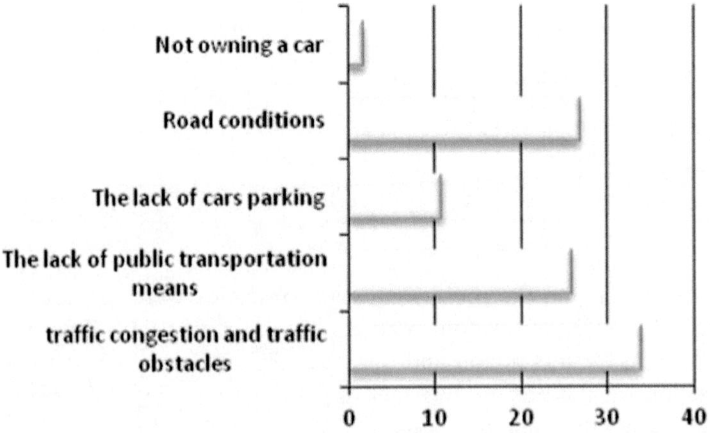

Fig. 5 The obstacles down to the final destination

Table 5 Infrastructure assessment for regional roads linked with Dujail

Degree	Condition of Dujail road network %	Condition of regional roads %	Traffic sign condition %	Road furniture condition %
Excellent	1	0	0	0
Very good	3	4	0	0
Good	18	9	3	1
Medium	31	36	27	15
Poor	47	51	70	84

As it turns out, the residents' assessment of the transportation network infrastructure in Dujail city ranged among low quality by 47%, medium 31%, and good 18%, Fig. 6. While assessing the road infrastructure throughout the region, it distributed among poorly by 51%, average construction status by 36%, and good by 9%. One of the fundamental reasons for the deterioration in the road network and Dujail residents' dissatisfaction is the lack of maintenance of this network for a long time and the region's recent military operations.

The traffic signs' status on the road was distributed among low quality by 70%, medium by 27%, and good by 3%. About the street furniture condition on these roads, the city's residents' opinions ranged among bad by 84%, medium quality by 15%, and good by 1%.

Perhaps the negligence that affected this sector after 2003 indicates the low quality of these services on this vital road network.

Traffic Accident. One of the most important indicators of the efficiency of the regional transport network is the decrease in traffic accidents that occur on the network. Accident rates often increase with private vehicles' use compared to public

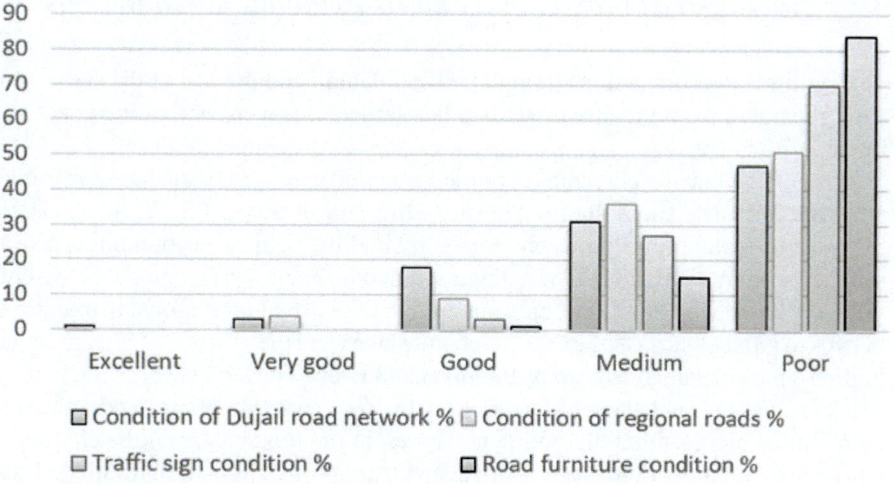

Fig. 6 Infrastructure assessment for regional roads linked with Dujail

Table 6 Traffic accidents

Type of car	Collision %	Run over %	Rollover %
Private car	67	74	71
Minibus	6	4	3
Truck	22	17	20
HOV	6	5	7
Agricultural puller	0	0	0
Total	100	100	100

transport users' accidents, as shown in Table 6. Accidents vary in terms of the level of damage, from material damage to severe and minor injuries or fatal accidents.

It is noted that the percentage of collisions caused by private cars reached 67%, while in the rest of the transport modes, it ranged between 22% for trucks 6% for minibusses and HOV.

The accidents of being run over by private cars reached 74%, while in the other transport modes, it ranged between 17% for trucks, to 5% for HOV and 4% for minibusses.

While the percentage of rollover accidents for private cars was 71%, and in other transport modes ranged between 20% for trucks, 7% for HOV, and 3% for minibusses.

It is noticed from the previous table that there are slightly high rates of accidents for trucks; this is because the roads of this region are the commercial link between Iraq and Turkey in which a high percentage of HOVs pass. There are also many accidents related to trucks, which are a common means of transportation due to the nature of agricultural use in the region.

3.2 *The Link Between Dujail and Its Regional Surroundings*

To determine the connection strength between Dujail and the rest of the settlements, trips generated from the city towards other settlements were determined, as shown in Table 7 and Fig. 7.

It is noticed that the percentages of the generated trips were distributed as follows:

Trips Towards the Villages Surrounding Dujail. From Fig. 8, it can be seen the most important links between Dujail city and its vicinity surroundings, near the villages scattered around it. The village of Sheikh Ibrahim represented the essential village with which Dujail residents interacted, with 3% of all regional trips and 31% of trips to villages close to the city. In second place were the villages of Sheikh Jameel, Sajllah, and Miznah, thirdly were Jolodiyat and Halif. The largest percentage of trips between the city and these villages are trips for work, shopping, and sometimes recreational, the fact that these villages represent the food basket of the city and the workplace of many of the city's residents who are proficient in agriculture and some of the crafts associated with it. As for the recreational aspect, the presence of farms and orchards is an outlet for vacation and recreation days for Dujail's residents.

Table 7 Regional interconnection between Dujail city and surroundings settlements

		Number of trips								
	Trip purpose	Work	Shopping	Education	Social & recreation	Health	Religious	Other	Total trips	%
Villages	Sheikh Ibrahim	60	20	6	22	2	12	15	137	3
	Sheikh Jameel	38	8	3	20	1	5	9	84	2
	Sajllah	45	14	4	32	0	6	11	112	2
	Jolodiyat	36	12	0	4	0	5	10	67	1
	Miznah	32	16	1	28	1	7	13	98	1
	Halif	24	6	1	12	1	7	8	59	1
Districts	Samaraa	215	50	103	40	20	100	40	568	12
	Tozkhormato	24	28	8	12	11	5	12	100	2
	Tikrit	307	52	140	26	72	5	65	667	14
	Beiji	32	16	5	20	4	0	17	94	2
	Balad	248	56	26	55	8	110	68	571	12
	Shorqat	20	4	0	10	0	0	20	54	1
	Dour	29	20	3	12	5	12	16	97	2
Governorates	Baghdad	493	105	78	47	105	55	84	967	20
	Ninawa	20	12	20	16	16	0	12	96	2
	Kerkuk	63	24	48	24	36	12	12	219	4
	Diyala	84	24	36	24	8	8	20	204	4
	Erbil	168	55	16	75	90	10	35	449	9
	Sulaimaniya	97	38	8	47	65	0	30	285	6
	Total trips	2035	560	506	526	445	359	497	4928	100
	%	41	11	10	11	9	7	10	100	

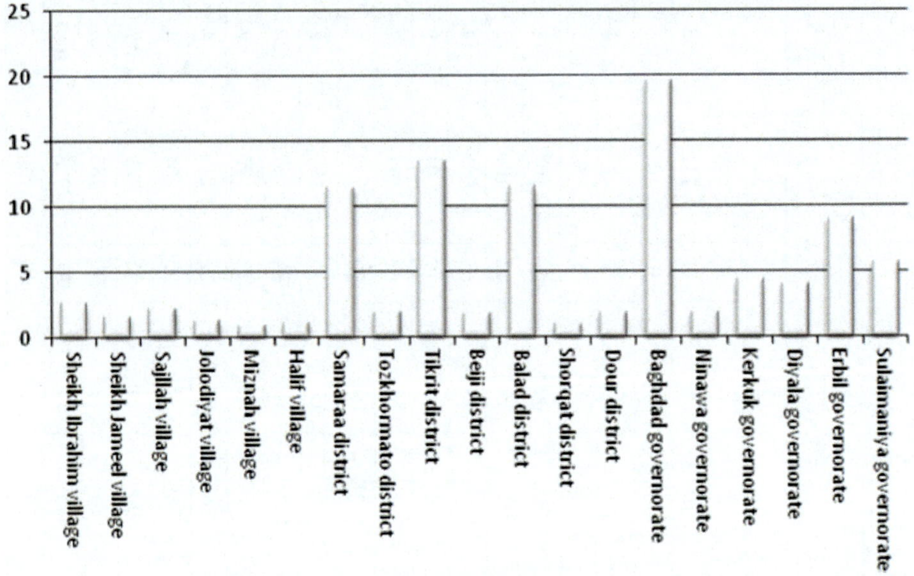

Fig. 7 Regional interconnection between Dujail city and surroundings settlements

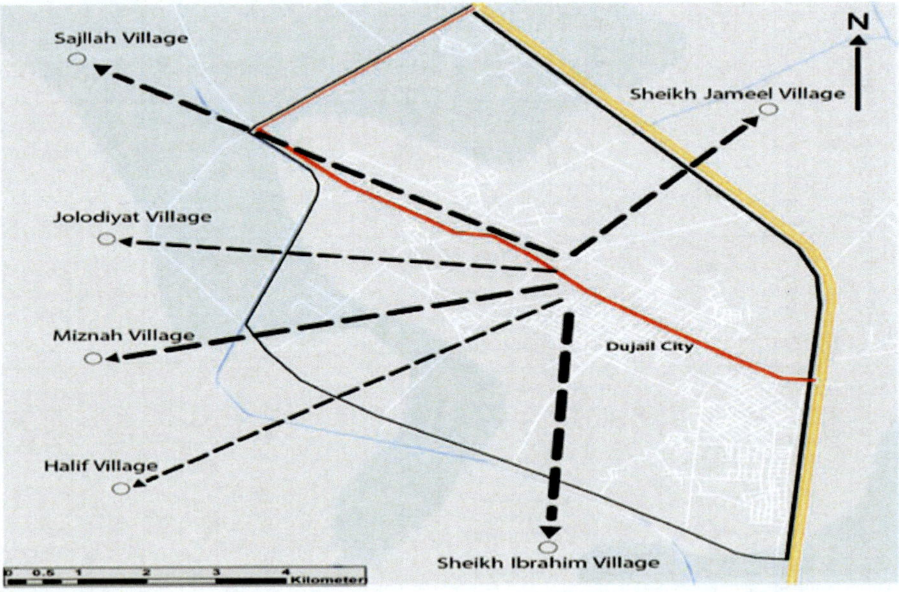

Fig. 8 Usual lines of desire for the trips of Dujail residents towards neighboring villages

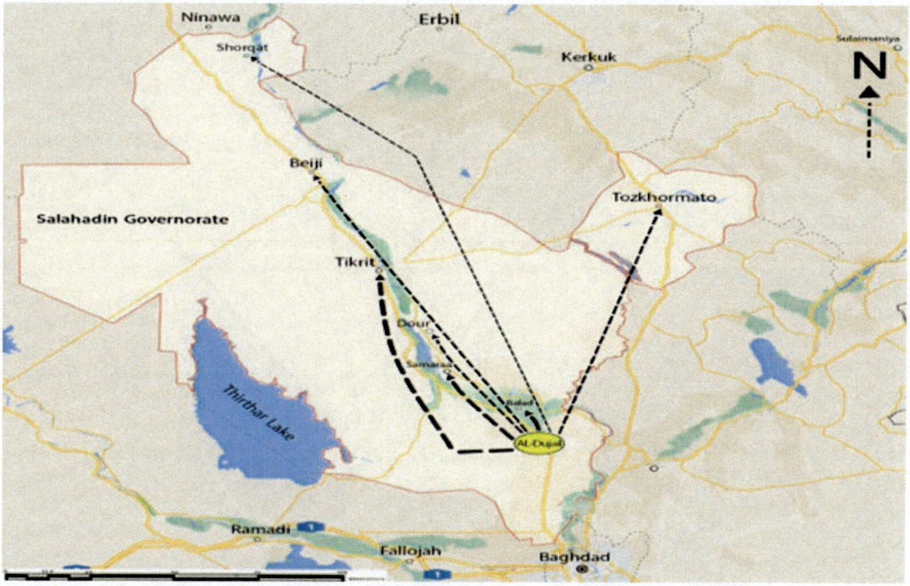

Fig. 9 Usual lines of desire for the trips of Dujail residents towards SG districts

Trips Towards the Districts of SG. As expected, most considerable interaction was recorded between Dujail city and Tikrit's district (the center of SG) in terms of the number of trips by 14%, Fig. 9. As the district of Tikrit represents the service and administrative center of SG and the University of Tikrit and some essential educational and services centers, as the residents mean it to complete their administrative transactions or shopping (wholesale markets). The district also includes the largest governmental university in the governorate, which includes most of the scientific and humanitarian specialties and the presence of a general hospital for all medical specialties. Therefore, we notice a higher concentration in business trips, shopping, education, and health than the rest of the districts of SG.

As for Samarra's district, there is a governmental university with a limited scientific and humanitarian specialty, a general hospital with specific specializations as well, simple wholesale markets, and religious shrines and archaeological sites. Therefore, we notice that the trips concentrated on work, shopping, education, recreation, and religious purposes.

About the district of Balad and by its proximity to Dujail city (less than 25 km), the trips have focused on work, as many of Dujail's residents have private businesses there. In addition to technical institutes' presence in it, which constitutes an educational attraction for Dujail's residents, religious shrines, and orchard areas, the trips concentrated among work, shopping, education, recreational and religious.

As for the other governorate districts, the percentage of its attraction to Dujail's residents was low due to its distance from the city and the availability of most services in the nearest settlements, which meet most of the requirements of Dujail city.

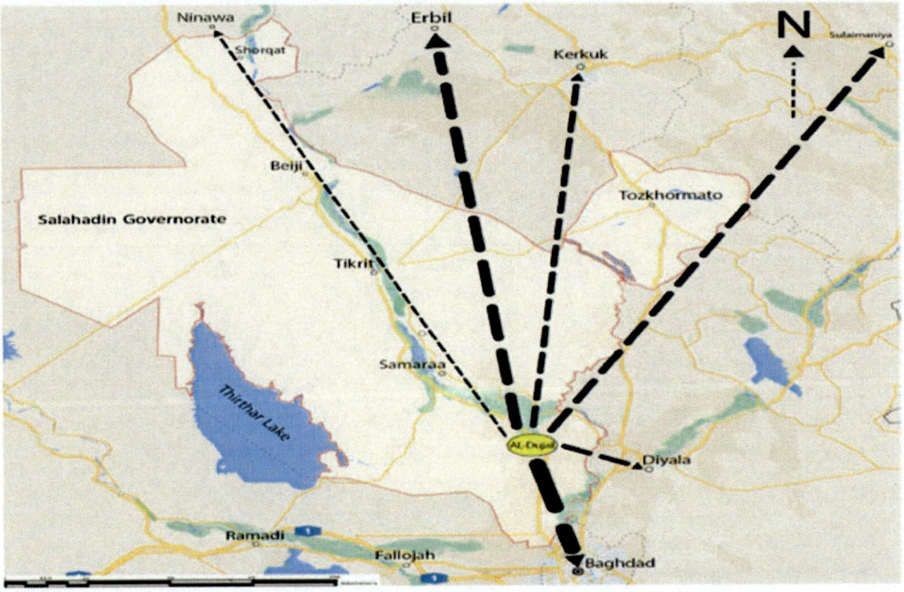

Fig. 10 Usual lines of desire for the trips of Dujail residents towards nearby provinces

Trips Toward Other Governorates. The questionnaire results indicated a concentration of trips towards Baghdad city capital of Iraq for most of the purposes. At a percentage, it reached 20% from all regional trips made by Dujail's residents, Fig. 10.

The nearness of Dujail city to Baghdad, relative to most SG districts, is the main factor contributing to Baghdad's city's concentration of trips if known that the volume and diversity of services are much better than in the rest of the regions. As the purposes of the trips to Baghdad varied among work, shopping, education, recreational, health, and religious, to varying percentages.

As for the Erbil governorate, its percentage reached 9% of the city's residents' total trips. While Sulaimaniya recorded 6% of the total trips, Kirkuk and Diyala 4% for each, and Nineveh only 2%.

The governorates of Erbil and Sulaimaniya are distinguished by their mountainous tourist character, representing an attraction for most of the Iraqi population in the summer season. The two governorates above also have a political and security stability that encouraged diversification of foreign investments in health and tourism and the presence of diplomatic missions, which increased their relative importance to the population of the study area.

4 Conclusion

The research results showed an excellent interaction of the city with the surrounding region due to the administrative, economic, educational, and health links. This interaction led to tremendous pressure on the road network leading to the settlements located in the region of Dujail city, and consequently, the deterioration of its infrastructure. The distance factor has a strong influence on the interconnectedness between Dujail and its region, as well as other functional influences, as evidenced by the relationship of Dujail with cities such as Baghdad and Balad.

The most considerable percentage of trips generated from the city of Dujail were heading towards Baghdad, which attracted 20% of the trips of Dujail residents. While Tikrit, Samarra, and Balad's districts came next in terms of intensity of interaction, at a third level came the governorates of Erbil and Sulaymaniyah.

The negligence of this region's roads and their failure to maintain them for nearly two decades and the military operations that concentrated in them led to the deterioration of the transport network there, thus increasing traffic accidents. The scarcity and inefficiency of public transportation in this region have increased private cars' importance as an alternative to transportation in this region.

Acknowledgements To those who work for others, to those who serve scholars, to Institute of International Education (IIE), the work you do will remain for eternally.

References

1. Riemann D (2013) Factors influencing regional transportation planning. A dissertation submitted for the degree of Master of Urban Planning, Design, and Development, Cleveland State University, The USA (2013)
2. Karner AA (2012) Transportation Planning and Regional equity: history, policy, and practice. A dissertation submitted for the degree of doctor of philosophy in Civil and Environmental Engineering, University of California, Davis, The USA (2012)
3. Alrawi F, Alrawi M (2009) Regional Development Under the organization of spatial structure of human settlements. J Alustath 96:405–430
4. Akpan US (2014) Impact of regional road infrastructure improvement on intra-regional trade in ECOWAS. Afr Dev Rev 26(S1):64–76
5. Loo BPY (1999) Development of a regional transport Infrastructure: Some lessons from the Zhujiang Delta, Guangdong. China. J Trans Geogr 7:43–63
6. Rodrigue JP (2020) The geography of transport system, 5th edn. Routledge, New York
7. Załoga E, Milewski D (2013) The impact of transport on regional development. Research Papers Of The Wroclaw University Of Economics/Prace Naukowe Uniwersytetu Ekonomicznego We Wroclawiu 286:71–78
8. Merkisz-Guranowska A, Bienczak M, Kicinski M, Zmuda-Trzebiatowski P (2016) Location of airports-selected quantitative methods. LogForum 12(3):283–295
9. Sierpiński G, Macioszek E (2020) Equalising the levels of electromobility implementation in cities. In: Mikulski J (ed) Research and the future of telematics, communications in computer and information science, vol 1289. Springer, Heidelberg, pp 165–176

10. Macioszek E, Sierpiński G (2020) Charging stations for electric vehicles - current situation in Poland. In: Mikulski J (ed) Research and the future of telematics, communications in computer and information science, vol 1289. Springer, Heidelberg, pp 124–137
11. Galińska B (2020) MCDM as the tool of intelligent decision making in transport. Case study analysis. In: Sierpiński, G. (ed.) Advances in intelligent systems and computing, vol 1091: Smart and Green Solutions for Transport Systems. Springer, pp 67–79
12. Tahoe Regional Planning Agency: Linking Tahoe Regional Transportation Plan. Technical report (2017)
13. Vallance P, Norman J (2019) A time of unprecedented change in the transport system. Technical report, Government Office for Science
14. Sierpiński G, Staniek M (2016) Education by access to visual information – methodology of moulding behaviour based on international research project experiences. In: Gómez Chova L, López Martínez A, Candel Torres I (2016) ICERI2016 Proceedings, 9th International conference of education, research and innovation, 14–16 November 2016, pp 6724–6729. IATED Academy, Seville, Spain. ISBN: 978-84-617-5895-1
15. Fengjun J, Jinxue D, Jiao′e W, Dong L, Chengjin W (2012) Transportation development transition in China. Chinese Geogr Sci 22(3):319–333
16. Li T, Cao X, Yang W (2016) Measuring the efficiency of regional integrated transport in China: data envelopment analysis. Periodica Polytechnica Transportation Engineering. 44(1):23–34
17. Closs DJ, Bolumole YA (2015) Transportation's role in economic development and regional supply chain Hubs. Transp J 54(1):33–54
18. Ivanova O (2003) The Role of Transport infrastructure in regional economic development. Technical report, The Institute of Transport economics
19. Gauthier HL (1970) Geography, Transportation, and regional development. Econ Geogr 4(4):612–619

Assessment Methods of Flexibility: A Systematic Overview of Land Transportation Systems

Simon Nagy and Csaba Csiszár

1 Introduction

Flexibility is an important characteristic of systems. In manufacturing, flexibility is a well-researched, well-developed field, which holds great potential, even after decades of intensive interest. Review of flexibility supports transportation system development significantly. Flexibility related literature, regarding transportation systems usually focus on resilience. Resilience is defined as resilience of infrastructure and services against weather, crises, nature catastrophes. Therefore, we identified our research direction (and a research gap) in defining flexibility of transport systems. We illustrated the steps of this research on Fig. 1.

Main steps of our research:

1. *Research questions*, as assessment-, key elements- and key development tools of flexibility regarding transportation systems.
2. *Aim and scope* is to review and analyse system flexibility, based on flexible manufacturing systems (FMS) and apply the concepts of FMS to transportation systems.
3. *The systematic review of literature,* divided into FMS-related and transportation system related assessments of flexibility. We compared and assessed the characteristics of flexibility.

S. Nagy (✉) · C. Csiszár
Faculty of Transportation Engineering and Vehicle Engineering, Department of Transport Technology and Economics, Budapest University of Technology and Economics, Budapest, Hungary
e-mail: nagy.simon@kjk.bme.hu

C. Csiszár
e-mail: csiszar.csaba@kjk.bme.hu

G. Sierpiński and E. Macioszek (eds.), *Decision Support Methods in Modern Transportation Systems and Networks*, Lecture Notes in Networks and Systems 208,
https://doi.org/10.1007/978-3-030-71771-1_3

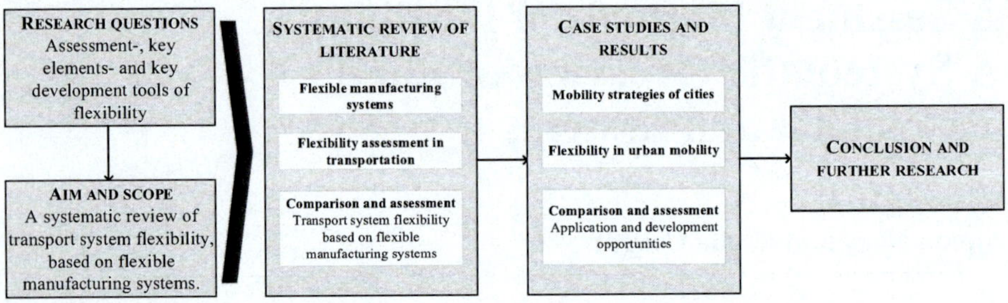

Fig. 1 Steps of the research

4. *Case studies and results,* in which we reviewed four city mobility plans, uncovered flexibility related terms, development aims and trends and finally reviewed the key development areas and opportunities.
5. *Conclusion and further research,* where we concluded the reviewed literature, as well as the city mobility plans in order to answer our research questions, and pointed out future research and development directions.

We reviewed 46 papers and five city mobility plans. We illustrated the reviewed papers on Fig. 2.

We aimed to introduce the fundamental literature, as well as the most recent ones. After the systematic overview, we discussed the results and defined development directions in order to increase the incorporation of flexibility aspects in city mobility plans. We found, that increasing flexibility has a great impact on increasing the overall performance of transport system development.

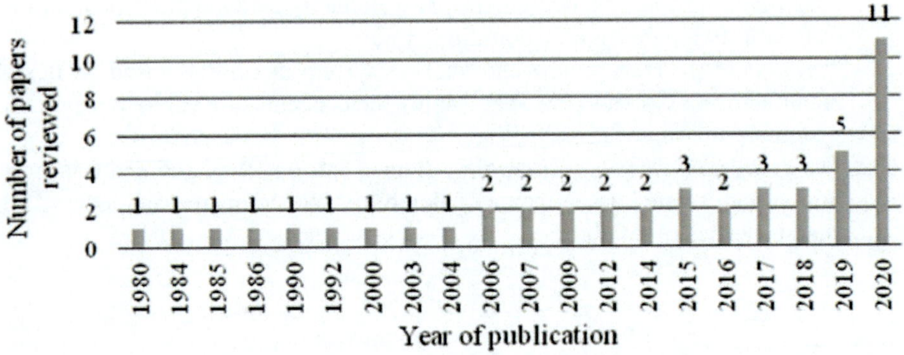

Fig. 2 The reviewed literature (according to year of publication)

2 Flexibility and Resilience in Literature

Flexibility is a crucial characteristic of any technological system due to their vulnerability. We review two main terms: flexibility and resilience. First, we discuss the flexibility related literature and conceptual basis of manufacturing systems. Then, we review the flexibility and resilience of transportation systems. Flexibility and resilience are related, multidimensional terms. Different dimensions are considered in different environments. Flexible operations and the ability of stable performance are the key challenges recently.

2.1 *Flexibility of Manufacturing Systems*

Flexible manufacturing system (FMS) is a production method, which is able to adapt to changes in type and quantity of production. A wide spectrum of literature is dealing with FMS since the early 80 s [1–5]. FMS is controlled by computers, and has some level of automation. Automated material handling devices, system monitoring processes and programmable transportation networks are identified as elements of FMS. The automation relies on numerical control techniques. Research interests according to FMS are design-, scheduling-, operation- and performance analysis and optimization. We summarized the different elements of flexibility in FMS in Table 1.

In Table 1., the abilities refer to the requirement towards a successful FMS. A higher level of automation and computer control increases the flexibility. Routing is one key element of FMS. Flexible routing is achieved by providing separate paths between machines or using a common material handling device, in which all parts are passed.

Table 1 Elements of flexibility related to FMS

Element	Ability
Machine flexibility	Making changes required to produce a given set of part types
Process flexibility	To produce a given set of part types each possibly using different materials
Product flexibility	To changeover to produce a set of products economically
Routing flexibility	To handle breakdowns and to continue producing the given set of part types
Volume flexibility	To operate an FMS profitably at different production volumes
Expansion flexibility	The capability of building a system and expanding it as needed easily and modularly
Operation flexibility	The ability to interchange the ordering of several operations of each part type
Production flexibility	The universe of part types, that the FMS can produce

Table 2 Recent research in FMS

Research direction	Related literature
Operations scheduling and optimization of FMS	[6, 7]
Facility layout design in order to support automation and FMS	[8–10]
Energy flexibility and efficiency of manufacturing systems	[11–13]
Maintenance optimization of automated manufacturing systems	[14, 15]

Though the definition of FMS was developed in the 80's, modern research still addresses flexibility. We concluded recent research directions with related literature in Table 2.

Scheduling of FMS have become a key challenge. The determination of palletizing for an upcoming schedule period is one key question for an economical and efficient operation. This includes pallet input sequency and routing, as well as pallet process sequency. Robust scheduling methods become more and more important. Robust scheduling means a suboptimum scheduling scheme, which has some level of resistance for stochastic disturbances (e.g. minor random machine fails).

A well-developed layout reduces the movement of workers, transportation devices, and materials. Layout problems are concerned with determining the positions of equipment and machinery in a facility. Aisle types, areas, borders, and doors are elements of infrastructure in a facility. Global optimization is an instrument of layout design. Machine layout types, machine loops and connections are principal research directions. Finally, with the automation of small logistics vehicles, facility layout faces new challenges.

Smart grid, a communication network, which handles information about energy consumption and optimizes efficiency, appears in the field of manufacturing as well. Applications, based on renewable energy sources, research towards self-sufficient facilities are key questions of research. Energy flexibility analysis and usage of integrated energy storage units is applied in facilities. Energy oriented strategies of production planning are emerging.

Lastly, with the increasing diversity of products and the appearance of many variants of part types, maintenance of FMS requires new techniques and solutions. In recent decades maintenance have developed. Opportunistic-, status-dependent, preventive- and reliability-based maintenance are just a few examples of different strategies. Application of advanced methods, learning algorithms, optimization of maintenance scheduling are emerging fields.

In conclusion, most modern research directions can be interpreted as optimization tasks. Optimization is focused on multiple things, e.g. operation, flow, design etc. Main tendencies in manufacturing system development are identified in Fig. 3.

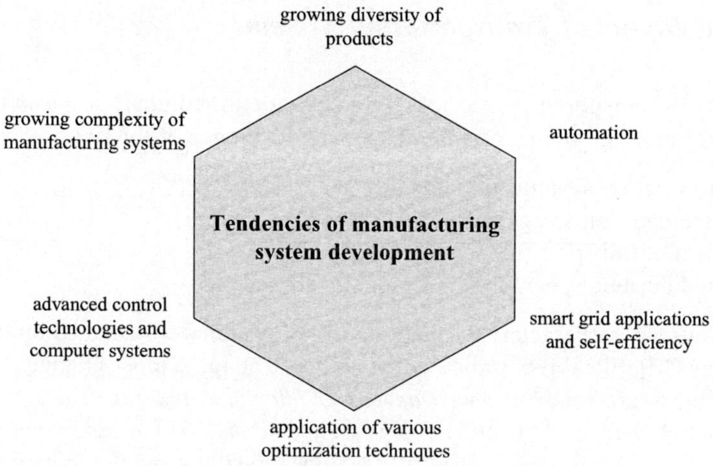

Fig. 3 Tendencies of manufacturing system development

Systems and parts are becoming more and more complex. The applied communication and computer technologies, with them the level of automation emerges. Based on the tendencies, we illustrated the requirements on Fig. 4.

Reliability means reliable operation of machines (low failure rate). Financial and energy efficiencies reflect an economical operation. High production quality stands for low failure rates of output products. Not all the requirements can be achieved at the same time. Accordingly, optimization of manufacturing systems is a key challenge.

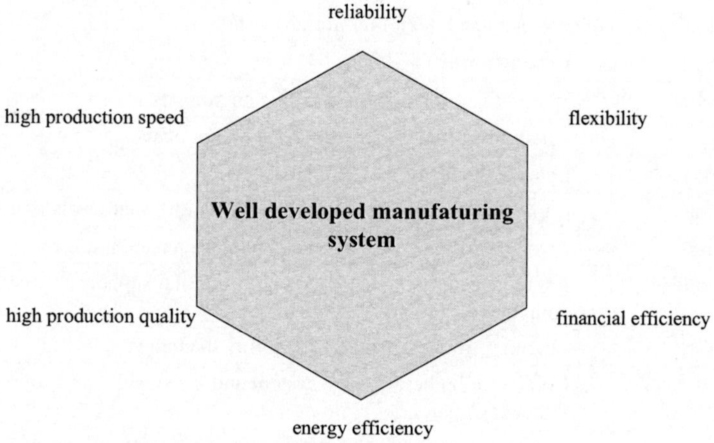

Fig. 4 Requirements towards a well-developed manufacturing system

2.2 Flexibility of Transportation Systems

Flexibility of transportation services have been identified many ways and has many variables. Literature focuses on flexibility and resilience of [16]

1. transportation systems in general,
2. different systems, e.g. rail, road,
3. subject of transport (passenger, freight),
4. spatial attributes, e.g. urban, regional, national etc.

Resilience is a characteristic that quantifies system performance under unusual conditions [17]. Flexibility indicates the ability to adapt to those changes. Resilience and flexibility are related: *a well-developed, flexible, and resilient system adapts to significant changes (e.g. changing passenger demand) but also resilient against certain impacts (e.g. weather).* A well-developed, resilient and flexible transportation system was described by the literature in many ways [18–20]. We summarized the main related terms in Table 3.

Resilience of systems is also approached from temporal side: static and dynamic resilience are identified [21, 22]. Static economic resilience refers to the ability of a system to maintain functionality when shocked from outside. Dynamic economic resilience adds the temporal perspective. In the dynamic approach, the speed of recovery is placed in focus.

Resilience and flexibility of railway networks are approached in the literature as well [23–25]. Research methodologies usually handles railway networks as graphs and elaborate strategies of service integration. Alternative path generation and bus

Table 3 Main terms of resiliency and flexibility in transportation

Term	Description
Redundancy	Alternative, functionally similar components
Diversity	Alternative, functionally different components
Efficiency	Energy efficiency, self-efficiency
Autonomous	Capability of operation without control, different levels
Robustness	A robust design so the system can resist outside impacts
Interdependency	Interconnected system components, internal support between components
Adaptability	Learning from experience, flexibility to change
Collaboration	Collaboration between the system and the stakeholders to improve efficiency/operations
Reliability	Probability, that the system functions adequately at a given time
Recoverability	Ability of a system to recover functionality
Rapidity	Term focusing on the speed of the recovering process
Resourcefulness	Availability of resources to support restoration of functionality
Preparedness	Prepare certain measures before disruption occurs

Table 4 Flexibility terms regarding transportation systems

Element	Description
Network flexibility	The ease, with which a network can adjust to changing circumstances and demands [30]
Capacity flexibility	The ability of a transportation system to accommodate variations or changes in traffic demand, while maintaining a satisfactory level of performance [31]
Redundancy	A measure of resiliency in transportation systems [32], it means alternative routes or ways to transport passengers or freight in a relation
Robustness	Ability, that a system can tolerate changes without adapting its' configuration, robustness is a requirement towards transportation systems, as e.g. robustness against environment [33, 34]

services linked to railway greatly increase flexibility and resilience. In dense urban areas, metro systems are special cases of railway. Peak-hour passenger rushes, high fluctuations are challenging aspects of metro systems.

Road networks are divided into physical infrastructure (road, bridges, crossings etc.), traffic and control systems. Flexibility and resilience are handled on those levels as well [26–28]. Technological aspects are the most discussed field, e.g. robust design methods, maintenance need analyses and prevention techniques of performance loss by design. With traffic on the roads though, user behaviour is a great limitation factor of elaboration of methods.

Resilience research is also divided, regarding subject of transport [29]. Resilience is addressed in the context of supply chains. Risk management, supply chain resilience strategies are identified. Enterprise activities (e.g. shipping plans) are key questions. Accordingly, enterprise resilience is analysed. Finally, the related infrastructure is taken into consideration (railway, road, airport).

We summarized the four most relevant and common terms regarding flexibility in Table 4.

Flexibility is in scope of simulations and optimizations, similar to FMSs. One key challenge, while developing transportation system is the heterogeneity. Spatial and city structure, demographic characteristics, stakeholders etc. are all heterogeneous [35, 36]. Inside city structure, urban land is divided to multiple areas (e.g. residential, commercial, industrial etc.), one of which is transportation network. Accordingly, flexibility does not mean the same for everyone. Different elements require different development methods.

We conclude that flexibility is less reviewed in the literature. Most of the reviewed papers focus on resilience. Elaborated methodologies are also mainly measuring resilience. Resilience is mostly associated with reliability, safety, and risk analysis. Methodologies are various; control systems, optimization, integration are key terms.

3 Conceptualizing Transportation System Flexibility

In this section, we have been described the main characteristics of flexible transportation systems. Our approach is based on FMS. Accordingly, we defined FMS to transportation and illustrated the definition on Fig. 5.

The machines of transportation are vehicles, carrying either passengers or goods. The product of a transportation system is a service: passenger transportation or freight transportation. Finally, the process is the transportation process, which is a complex term. Infrastructure (for transportation and information management) and companies (e.g. public transport companies, freight transport companies, vehicle sharing companies) are also part of transportation system. *Infrastructure is reviewed as not only rail and road but also information systems (software and hardware).*

The difference between transportation service and transportation process is decisive. *In economics, service means a transaction, during which no physical goods are transferred.* In the context of transportation, service means the transfer of passengers or goods from an origin to a destination.

Transportation process is a complex, engineering term. Processes are series of interrelated tasks, in which raw materials (inputs) are transformed into a product (output). Inputs are vehicles (with fuel, driver, legal contracts etc.), infrastructure (road and rail), supporting systems (e.g. ticket system of public transport) and subject of transportation (passengers, goods). The output of the transportation process is the transportation service.

3.1 *Defining Flexibility of Land Transportation Systems*

We define the flexibility elements of transportation systems, based on the eight elements of FMS on Fig. 6.

We identified four elements of flexibility, regarding transportation systems:

- vehicle flexibility, which stands for the change of vehicles, according to demand (e.g. rush-hour peaks, special traveller groups),

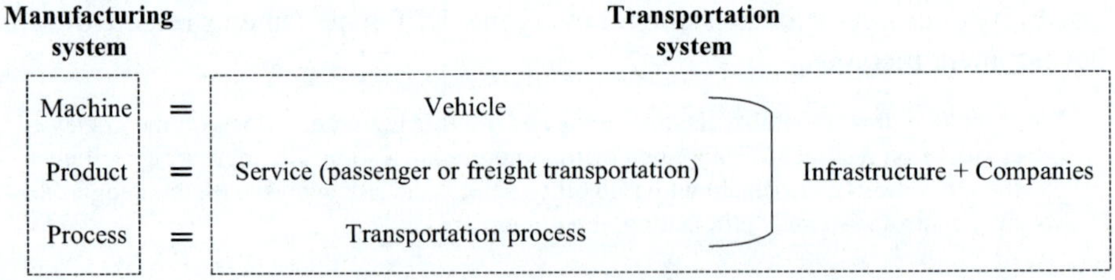

Fig. 5 The definition of transportation system based on manufacturing systems

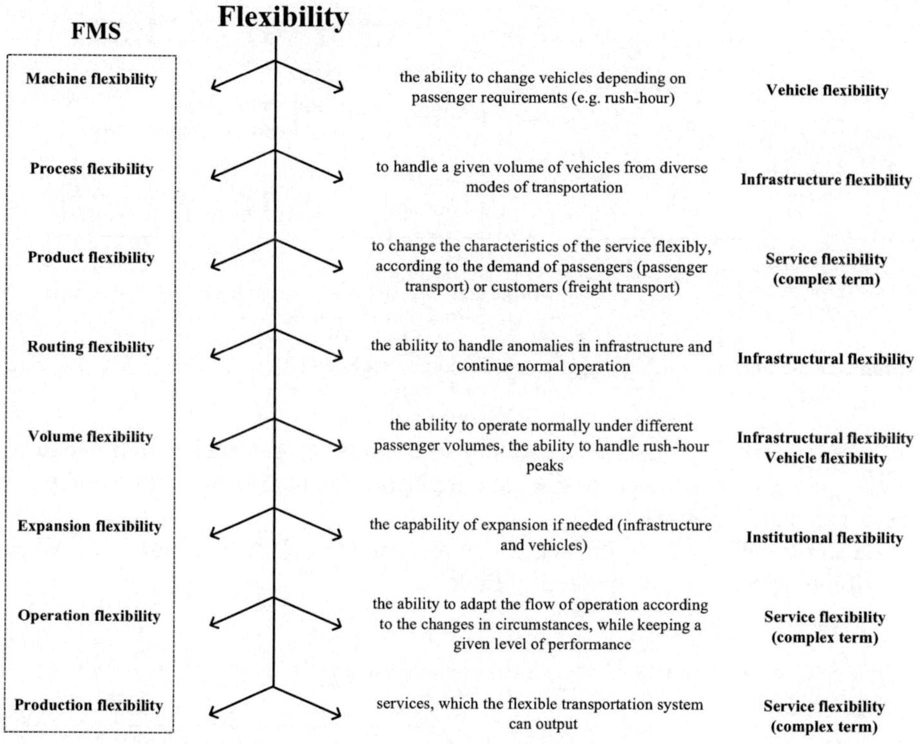

Fig. 6 Flexibility elements of land transportation, based on FMS

- infrastructure flexibility, which means, that the infrastructure can handle different volume of vehicles from diverse modes, as well as anomalies,
- service flexibility, which is complex term meaning (1) flexible service characteristics, e.g. price, routing, vehicles; (2) adaptation of operation for different circumstances; (3) various services in the system (opportunity of the citizens to choose),
- institutional flexibility, which reflects to the flexibility of stakeholder institutions and companies.

The four main flexibility elements include various 'sub-elements'. We summarized those in Fig. 7.

Vehicle flexibility F_v are identified with flexibility of vehicle types. These types are private vehicles, vehicles of citizens or private companies and public vehicles. Regarding infrastructural flexibility F_i physical infrastructure stands for rail and road. Informational infrastructure is hardware (e.g. stationary informational devices at roadside or stations; informational devices in the vehicles, personal informational devices) and software (e.g. applications, operational systems of devices etc.). Inside service flexibility F_s we differentiate passenger- and freight transportation

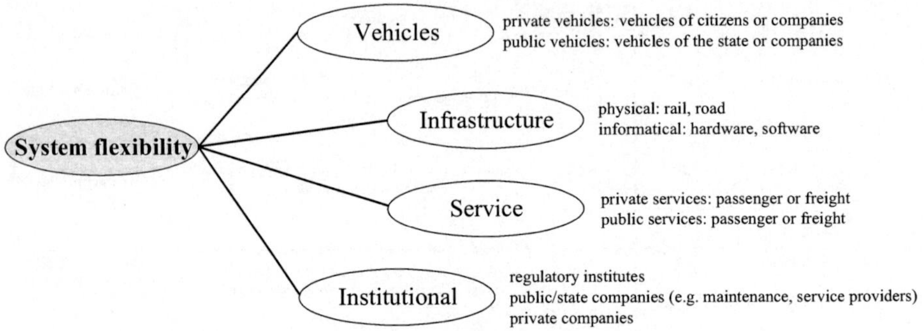

Fig. 7 Sub-elements of flexibility of land transportation systems

services. Finally, we identify regulatory institutions F_n e.g. transportation authorities), public/state institutes (e.g. public transport company) and private companies (e.g. car-sharing companies).

Accordingly, we have defined a complex, multi-stakeholder and -dimensional flexibility approach, summarized in Table 5.

Table 5 Concept of flexibility in land transportation systems

Element F_t	Sub-element F_f	Category F_k
Vehicle F_v	Private	Individual vehicles
		Company vehicles
	Public	Passenger transport
		Freight transport
Infrastructure F_i.	Physical	Rail (railway, stations)
		Road (roads, stations, hubs, parking spaces)
	Informatical	Hardware (devices at stations, on-board devices, personal devices)
		Software (applications, operating systems)
Service F_s	Passenger transport services	Public services
		Private services
	Freight transport services	Public services
		Private services
Institution F_{in}.	Regulatory institutes	Transport authorities, governments
	Public/state companies	Service providers (passenger and/or freight transport)
	Private companies	Service providers (passenger and/or freight transport)

3.2 *Information-Based Assessment of Flexibility*

Flexibility assessment methods usually use graph-based illustration and analysis [37, 38]. In the context of transportation systems, the illustration is rather complex. Infrastructure with links (rail, road, communication lines) and nodes (crossings, stations, info-communication technology (ICT) base-stations) can be illustrated as a graph, as well as a few other elements/sub-elements, e.g. public transport services.

Assessment methodologies, frameworks, as well as analysis- and illustration techniques have been moving towards ICT applications and advanced informatics, although keeping graphs and certain networks (e.g. Bayesian) as mathematical methods [39–41]. Information based methodologies are advantageous, as a transportation system is complex and changing dynamically. Great amount of data is present and play key role in establishing efficient development and operation. The concept of intelligent transportation system (ITS) has great potential in handling system assessment problems, like flexibility. Sustainable development is also recognized, as a multi-stakeholder process [42, 43]. We recommend to include citizens, companies etc. into the development process. Accordingly, through the inclusion a wide-range of ICTs are used (e.g. smart devices, web-based applications etc.)

Accordingly, we recommend using an information-based approach. We illustrated the information-based assessment process on Fig. 8.

On Fig. 8., we illustrate the data sources divided: on the left side, vehicles, infrastructure, services and institutions are illustrated inside white boxes, while on the right side, in the grey boxes the exact data sources. The data is collected and transmitted into a central database. The analysis is done, in respect of the collected data. Finally, we illustrated the results; these support system flexibility development.

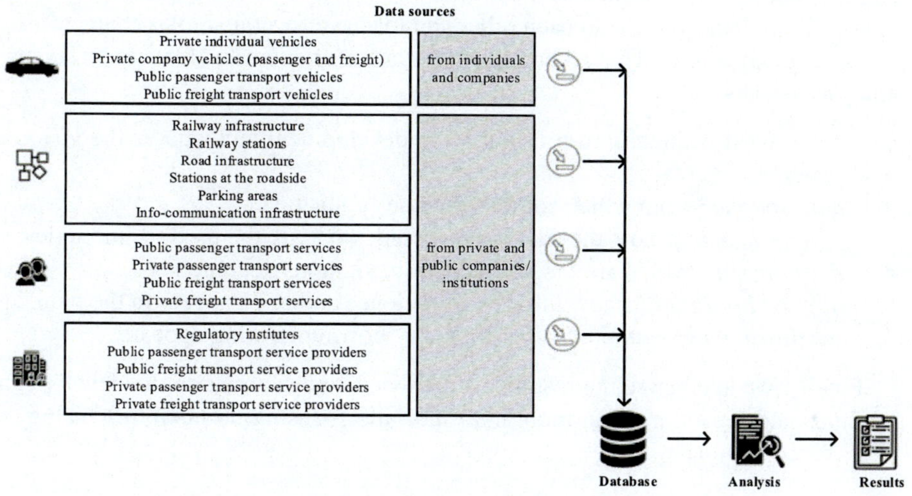

Fig. 8 Information-based assessment process of transport system flexibility

The structure of results is the key question to efficiently support flexibility development on the system level. We suggest indicator-driven results. This means preparing a set of indicators, with which the performance can be described constantly. The constant monitoring and indicator development on the system level require city level strategies.

Indicators are various and have been revealed by several research in multiple contexts (e.g. sustainability, resilience) [44–50]. Indicators are quantitative or qualitative; some issues (e.g. emission) are easily quantified. The common direction in indicators are the aims of a sustainable and efficient operation. Defining indicators for flexibility is a complex issue. We found, that both quantitative and qualitative indicators are needed. In an optimal case, we are always monitoring the current performance of an issue with indicators. Next to indicators, the analysis of connections has great potential. As we uncovered in our previous research [51] different modes require different models, approaches. Accordingly, we recommend the analysis of spatial connection for every mode.

Quantitative indicators are used with a discrete, marginal value; e.g. if the emission is lower than a given value, the systems' emission performance is good. Qualitative indicators require principles, policies, recommendations or references, and the progression must be tracked manually. The indicators are constantly monitored and summarized in a city mobility plan.

4 Case Studies and Results

In this section we review the city mobility plans of Vienna, Budapest, Brno, Warsaw and Prague [52–56]. We choose these cities, as (1) they have well-defined mobility plans, (2) are located close to each other and (3) have several similar characteristics (area, development level etc.). Our review is system-oriented and focuses on six main analysis aspects:

1. *aims:* what is the aim of transportation development and what is the vision of cities,
2. *data sources:* from which sources do cities collect data,
3. *data processing:* how the data is processed, what are the used technologies,
4. *development:* which are the key development fields, techniques
5. *applied informatics:* how information science and ICTs appear in the plan, and
6. *monitoring and continuous development:* how monitoring appears.

Finally, we synthetize the results of the review and summarize the development methods and trends in urban mobility, especially focusing on flexibility, using our FMS based approach.

4.1 City Mobility Plans: Vienna, Budapest, Brno, Warsaw and Prague

The aims are identified discretely in each city. First, there are visionary goals, which are general. Inside visionary goals, strategic goals are identified. Under the strategic goals, each city follows different goal establishment techniques. Generally, *the aim system of each city focuses on*

- *sustainability:* environmental and social aspects and externalities of transportation,
- *liveable city development:* minimizing the externalities, efficient resource usage, such as fuel, energy, public space,
- *cooperative development:* development of transportation system together with the remote regions and suburbs of the city,
- *modal split:* reducing the usage of private cars, while increasing the usage of public transport, pedestrian traffic, and cycling,
- *infrastructure design:* accessibility of transportation, efficient and fair use of public space,
- *safety and reliability:* increasing the safety of transportation, as well as developing a reliable and robust system.

We summarized the reviewed mobility plans in Table 6.

The main difference between the city mobility plans is data sources and centralization. Budapest follows a centralized practice. Data is collected, processed, and reported by the central development organization. Warsaw does not report the data sources nor the process of collection. We presume data collection is centralized. Brno, Prague, and Vienna however are rather decentralized. Sharing of data and activities appears on the strategic level. *Decentralization of development processes increases the flexibility of the development.*

Data processing techniques are identified in great detail regarding Budapest. Brno, Prague, and Vienna have been a bit different here as well. While Budapest differentiates statistical, engineering modelling, economical modelling and project management techniques, the data processing is centralized as well. Warsaw simplifies the processing into a SWOT analysis. We presume a wide range of engineering measurements have been done, as several maps and network illustrations justify the goals. Warsaw also follows a centralized practice. The other three cities greatly include citizens/passengers and their preferences. *Flexibility of data collection and processing is more significant in Vienna, Brno and Prague, as multiple sources are used, and the participants are included.*

Development ideas are similar in each city. Budapest includes a list of possible projects in detail. The projects are ranked and illustrated (maps, network schemes). The analysis of strategic goals, monitoring of projects also have high importance. Warsaw states main objectives and inside specific objectives. Objectives also divided by spatial units. Spatial units, as well as modes have several specific, well-defined policy principles. Brno and Prague evaluate indicators for each strategic goal, services

Table 6 The reviewed mobility plans according to flexibility

Category	City	Characteristic
Data sources	Budapest	Centralized
	Warsaw	Centralized
	Brno	Decentralized
	Prague	Decentralized
	Vienna	Decentralized
Data processing	Budapest	Centralized
	Warsaw	Centralized
	Brno	Decentralized
	Prague	Decentralized
	Vienna	Decentralized
Development	Budapest	Detailed list of projects and measures
	Warsaw	Objectives, policy principles for spatial units and modes
	Brno	Evaluation indicators, monitoring
	Prague	Evaluation indicators, monitoring
	Vienna	Soft measures, policies, recommendations
Applied informatics	Budapest	Measures and databases, centralized
	Warsaw	Applied informatics in mobility management and planning
	Brno	Personal devices, databases, ICTs in transport planning
	Prague	Personal devices, databases
	Vienna	Personal devices, databases
Monitoring and continuous development	Budapest	Complex indicator system, multiple levels
	Warsaw	Policy principals, centralized
	Brno	Indicators, strategic level
	Prague	Indicators and benchmarking system
	Vienna	Soft policies

measures, and target values. Vienna sets list of focus areas and policies, however, the indicators and monitoring systems are not identified discretely.

Applied informatics is described in all mobility plans. The centralized mobility plan of Budapest applies information systems and databases in the measurement- and development processes. Warsaw stresses the importance of system integration, which relies heavily on informatics. Furthermore, control systems, a high technical and management standard etc. have been identified. Accordingly, the mobility plan of Warsaw includes applied informatics in several ways/layers. Brno, Prague and Vienna rely greatly on personal devices, informatics in data collection, as well as informatics in data processing. Each mobility plan includes the idea of a digital city. Brno emphasizes the importance of ICTs on the visionary level: road communication network and quality of public resources.

Finally, the monitoring and continuous development have been done through indicators and constant monitoring quality management systems in Budapest, Brno and Prague as well. Budapest has the most complex indicator system. Indicators are identified in multiple levels: extensive/general, strategic, operative, activity, project. Brno has multiple indicators on the strategic level. Prague includes a benchmarking system next to the strategic indicators. Benchmarking have been done regarding competitive, similar cities, in advance to find good practices and development opportunities. In Vienna, as the mobility plan is softer, less attention is paid on continuous monitoring. The mobility plan of Warsaw has not been defined continuous development processes; however, the elaborated policy principals enable a centralized monitoring system.

In conclusion, we found that Budapest and Warsaw have the least flexible mobility plans. Budapest also has the highest level of resolution, with multiple goal levels, indicators, list of projects. Warsaw's mobility plan introduces a great level of detail, regarding modes and actions. Brno and Prague are similar, they have flexible and well-defined mobility plans. Brno have the advance of defining the importance of ICT in transport technology on the strategic level, while Prague advances with benchmarking. Vienna's mobility plan is a soft, vision and policy-oriented document. It has a well-defined goal system, which advances over the other three by phrasing and emphasizing the importance of public space usage.

5 Discussion

Through the systematic review of FMSs and transportation system flexibility, we identified the main characteristics of flexibility. FMSs have a well-developed, broad literature, from which the flexibility elements (and sub-elements) were extracted and defined to transportation. Transportation system flexibility on the other hand is not that researched. Papers usually focus on resiliency, robust design, resistance against weather and disasters etc.

Flexibility of transportation systems is a key question, as a well-designed, flexible system is efficient, robust, adaptive and serves a broad range of alternatives for passengers. Infrastructure, vehicles, services, and institutions are all part of the complex system of transportation. Accordingly, flexibility appears on many levels in many forms.

After the overview of city mobility plans, we found, that a great distinctive factor is the level of centralization. Centralized strategies (like Budapest and Warsaw) tend to serve every step of the development process in detail; however, the inclusion

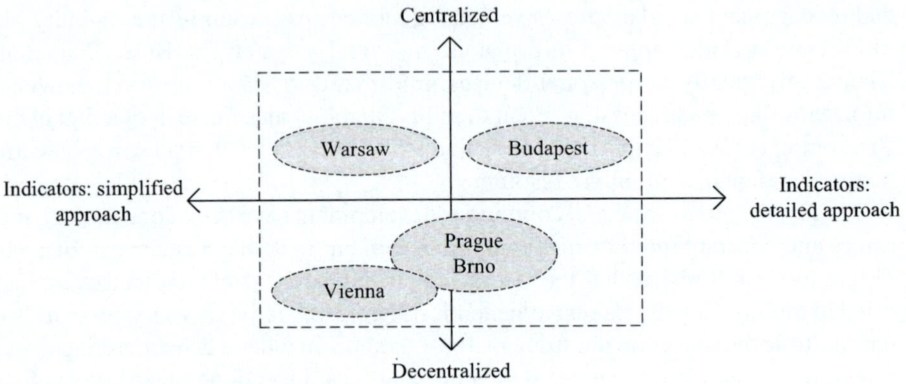

Fig. 9 The reviewed city mobility plans: centralization and indicator detailedness

of citizens and private companies is not significant. Decentralized mobility plans are including citizens, companies into the development process. Another difference is regarding indicators. Against Budapest's detailed approach, mobility plans of Brno and Prague are simplified, while Vienna and Warsaw are soft, policy-oriented. Accordingly, monitoring and constant development processes are different as well. We illustrated the reviewed city mobility plans, regarding centralization and indicator detailedness on Fig. 9.

In order to increase flexibility, we recommend

- the decentralization of data collection and processing, by the inclusion of citizens, passengers,
- establishing a central database, accessible by all stakeholders,
- finding the balance between an overly detailed indicator system and a soft, policy-based practice in order to track and monitor development progress,
- elaborating an inter-city or even an international benchmarking system, in order to seek best practices and develop in a more efficient way,
- prioritizing energy efficient and environmentally friendly modes, as well as ICT development, and
- applying information science at all levels of planning and development.

In sum, mobility plans do not need to include everything with the highest detail, and should be flexible in the way, that citizens and passengers could influence the development processes.

6 Conclusion

The main contribution of this chapter is a systematic overview of literature and city mobility plans in order to summarize and critically describe flexibility in transport systems. Flexibility is often researched together with resiliency. Accordingly, most of the literature focus on resilience against weather, crises, or nature catastrophes. Therefore, we identified the complex definition of flexibility as our research direction. We found, that

- flexibility in transportation is a multidimensional phrase,
- we assigned flexibility elements to vehicle-, infrastructure-, service- and institutional flexibility,
- we uncovered, that different elements of flexibility require different approaches in assessment, analysis, and presentation,
- accordingly, we identified an information-based assessment process, which can be applied to all elements of flexibility,

Secondly, we reviewed four city mobility plans, in order to uncover the level of flexibility and found, that

- centralization and decentralization are great distinctive factors of how flexible a system is, that is why we recommend increasing decentralization and strengthening cross-functional processes (between different stakeholders) and cooperation,
- we identified that mobility plans include several indicators; however, the level of detail (discreteness, linked goals, evaluation) differs greatly,
- we identified that only one of four cities define a benchmarking system in their mobility plans; we recommend evaluating a benchmarking system supported by applied informatics, in order to track comparable cities and reveal good practices,
- finally, we found, that city mobility plans are usually include flexibility 'under the surface' and most of our development recommendations aimed on structural and phrasing issues.

We uncovered, that flexibility of transportation systems, in the general term is not well-researched. Literature usually focuses on discrete, practical issues. Therefore, we identify great research potential in transportation system flexibility. Our future research will focus on mathematical phrasing of flexibility, modelling, assessing the flexibility of the different elements, as well as introducing the new concepts and methodologies to practice (urban- and transportation system development).

Acknowledgements The research was supported by the Ministry of Innovation and Technology NRDI Office within the framework of the Autonomous Systems National Laboratory Program.

References

1. Browne J, Dubois D, Rathmill K, Sethi SP, Stecke KE (1984) Classification of flexible manufacturing systems. The FMS Mag 2:114–117
2. Banaszak ZA, Krogh BH (1990) Deadlock avoidance in flexible manufacturing systems with concurrently competing process flows. IEEE Trans Robot Autom 6:724–734
3. Buzacott JA, Yao DD (1986) Flexible Manufacturing Systems: A Review of Analytical Models. Management Sci. 32:890–905

4. Buzacott JA, Shanthikumar JG (1980) Models for understanding flexible manufacturing systems. A I I E Trans 12:339–350
5. Kusiak A (1985) Flexible manufacturing systems: a structural approach. Int J Prod Res 23:1057–1073
6. Lee DK, Shin JH, Lee DH (2020) Operations scheduling for an advanced flexible manufacturing system with multi-fixturing pallets. Comput Ind Eng 144:
7. Wang Z, Pang CK, Ng TS (2019) Robust scheduling optimization for flexible manufacturing systems with replenishment under uncertain machine failure disruptions. Control Eng Pract 92:
8. Pourvaziri H, Pierreval H, Marian H (2020) Integrating facility layout design and aisle structure in manufacturing systems: Formulation and exact solution. Eur J Oper Res (available online)
9. Benderbal HH, Benyoucef L (2019) A new hybrid approach for machine layout design under family product evolution for reconfigurable manufacturing systems. IFAC-PapersOnLine 52:1379–1384
10. Gutta PR, Chintala VS, Manchoju RV, Charan V, Purohit R (2018) A review on facility layout design of an automated guided vehicle in flexible manufacturing system, Materials Todays: Proc. 5, pp 3381–3986
11. Schulze C, Blume S, Siemon L, Herrmann C, Thiede S (2019) Towards energy flexible and energy self-sufficient manufacturing systems. Procedia CIRP 81:683–688
12. Roth S, Stumpe L, Schmiegel B, Braunreuther S, Schilp J (2020) An optimization-based approach for the planning of energy flexible production processes with integrated energy storage scheduling. Procedia CIRP 88:258–264
13. Keller F, Schultz C, Simon P, Braunreuther S, Glasschröder J, Reinhart G (2017) Integration and interaction of energy flexible manufacturing systems within a smart grid. Procedia CIRP 61:416–421
14. Wocker M, Betz NK, Feuersänger C, Lindworsky A, Deuse J (2020) Unsupervised learning for opportunistic maintenance optimization in flexible manufacturing systems. Procedia CIRP 93:1025–1030
15. Celen M, Djurdjanovic D (2012) Operation-dependent maintenance scheduling in flexible manufacturing systems. CIRP J Manuf Sci Tech 5:296–308
16. Goncalves LAPJ, Ribeiro PJG (2020) Resilience of urban transportation systems. Concept, characteristics, and methods. J Transp Geogr 85:102727
17. Murray-Tuite PM (2006) A comparison of transportation network resilience under simulated system optimum and user equilibrium conditions. In. Proceedings of the 2006 winter simulation conference, pp 1398–1405. IEEE Press, Monterey, CA
18. Godschalk DR (2003) Urban hazard mitigation: creating resilient cities. Nat Hazard Rev 4:136–143
19. Mattsson LG, Jenelius E (2015) Vulnerability and resilience of transport systems—A discussion of recent research. Transp Res Part A: Policy Pract 81:16–34
20. Wan C, Yang Z, Zhang D, Yan X, Fan S (2018) Resilience in transportation systems: a systematic review and future directions. Transp Rev 4:479–498
21. Pant R, Barker K, Zobel CW (2014) Static and dynamic metrics of economic resilience for interdependent infrastructure and industry sectors. Reliab Eng Syst Safety 125:92–102
22. Rose A (2007) Economic resilience to natural and man-made disasters: multidisciplinary origins and contextual dimensions. Environ Hazards 7:383–398
23. Jin JG, Tang LC, Sun L, Lee DH (2014) Enhancing metro network resilience via localized integration with bus services. Transp Res Part E 63:17–30
24. Roanes-Lozano E, Laita LM, Roanes-Macías E, Wester MJ, Ruiz-Lozano J, Roncero C (2009) Evolution of railway network flexibility: the Spanish broad gauge case. Math Comput Simul 79:2317–2332
25. D'Lima M, Medda F (2015) A new measure of resilience: an application to the London underground. Transp Res Part A 80:35–46
26. Nogal M, Honfi D (2019) Assessment of road traffic resilience assuming stochastic user behaviour. Reliab Eng Syst Safety 185:72–83

27. Ouyang M, Wang Z (2015) Resilience assessment of interdependent infrastructure systems: With a focus on joint restoration modeling and analysis. Reliab Eng Syst Safety 141:74–82
28. Nogal M, O'Connor A, Caulfield B, Martinez-Pastor B (2016) Resilience of traffic networks: From perturbation to recovery via a dynamic restricted equilibrium model. Reliab Eng Syst Safety 156:84–96
29. Ta C, Goodchild AV, Pitera K (2009) Structuring a definition of resilience for the freight transportation system. Transp Res Rec J Transp Res Board 2097:19–25
30. Feitelson E, Salomon I (2000) The implications of differential network flexibility for spatial structures. Transp Res Part A 34:459–479
31. Morlok EK, Chang DJ (2004) Measuring capacity flexibility of a transportation system. Transp Res Part A: Policy Pract 38:405–420
32. Xu X, Chen A, Jansuwan S, Yang C, Ryu S (2018) Transportation network redundancy: complementary measures and computational methods. Transp Res Part B: Method 114:68–85
33. Henry D, Ramirez-Marquez JE (2012) Generic metrics and quantitative approaches for system resilience as a function of time. Reliab Eng Syst Safety 99:114–122
34. Zhang W, Wang N (2016) Resilience-based risk mitigation for road networks. Struct Saf 62:57–65
35. Fonseca JA, Estévez-Mauriz L, Forgaci C, Björling N (2017) Spatial heterogeneity for environmental performance and resilient behavior in energy and transportation systems. Comput Environ Urban Syst 62:136–145
36. Cadenasso ML, Pickett ST, Schwarz K (2007) Spatial heterogeneity in urban ecosystems: reconceptualizing land cover and a framework for classification. Front Ecol Environ 5:80–88
37. Chandra P, Tombak MM (1992) Models for the evaluation of routing and machine flexibility. Eur J Oper Res 60:156–165
38. Kaschel H, Sánchez y Bernal LM (2006) Importance of flexibility in manufacturing systems. Int J Comput Commun Control 1:53–60
39. Koc E, Cetiner B, Rose A, Soibelman L, Taciroglu E, Wie D (2020) CRAFT: comprehensive resilience assessment frameworks for transportation systems in urban areas. Adv Eng Inf 46:
40. Tang J, Heinimann H, Han K, Luo H, Zhong B (2020) Evaluating resilience in urban transportation systems for sustainability: a systems-based Bayesian network model. Transp Res Part C: Emerging Technol 121:
41. Richter A, Löwner M-O, Rüdiger E, Scholz M (2020) Towards an integrated urban development considering novel intelligent transportation systems: urban development considering novel transport. Tech. Forecasting and Social Change 155:
42. Myrovali G, Morfoulaki M, Vassilantonakis BM, Mpoutovinas A, Kotoula KM (2020) Travelers-led innovation in sustainable urban mobility plans. Periodica Polytechnica Trans Eng 48:126–132
43. Sierpiński G (2013) Revision of the Modal Split of Traffic Model. In: Mikulski J (ed) Transport systems telematics 2013, vol 395. Activities of transport telematics, communications in computer and information science. Springer, Heidelberg, pp 338–345
44. Leobons CM, Campos VBG, Bandeira RAM (2019) Assessing urban transportation systems resilience: a proposal of indicators. Trans Res Procedia 37:322–329
45. Naganathan H, Chong WK (2017) Evaluation of state sustainable transportation performances (SSTP) using sustainable indicators. Sust Cities Soc 35:799–815
46. Yang Z, Poo MCP, Galatioto F, Dimitru D, Qu Z, Rushton C, Lee PT, Guan B, Woodward N (2020) Key Green Performance Indicators (KGPIs) for vehicle cleanliness evaluation: a buyer choice. Transp Res Part D: Transp Environ 87:102505
47. Danchuk V, Bakulich O, Svatko V, Bieliatynskyi A (2021) Simulation of traffic flows optimization in road networks using electrical analogue model. Adv Intell Syst Comput 1258:238–254
48. Pashkevich A, Nõmmik A, Antov D (2017) Competitiveness analysis of regional airports based on location planning models: the case study of Finland. In: 21st International scientific conference transport mean. Kaunas University of Technology, Kaunas, pp 754–761

49. Macioszek E, Sierpiński G (2020) Charging stations for electric vehicles—current situation in Poland. In: Mikulski J (ed) Research and the future of telematics, communications in computer and information science, vol 1289. Springer, Heidelberg, pp 124–137
50. Macioszek E, Sierpiński G, Staniek M (2017) Analysis of trends in development of freight transport logistics using the example of Silesian Province (Poland) – a case study. Transp Res Procedia 27:388–395
51. Nagy S, Csiszár Cs (2020) Analysis of ride-sharing based on Newton's gravity model. In: 2020 Smart City symposium Prague (SCSP), pp 1–6
52. Urban Mobility Plan Vienna: short report. http://sump-network.eu/fileadmin/user_upload/SUMPs/Vienna_SUMP_summary_EN.pdf
53. Budapest Mobility Plan 2030. https://budapest.hu/Lapok/2019/budapesti-mobilitasi-terv-2030.aspx
54. Brno Mobility Plan. http://www.mobilitabrno.cz/data_files/ostatni/brno-mobility-plan-eng.pdf
55. The sustainable mobility plan for Prague and its suburbs. https://poladprahu.cz/wp-content/uploads/2019/11/Mobility_Plan-Brochure_EN.pdf
56. The transportation system of Warsaw: sustainable development strategy. http://www.transport.um.warszawa.pl/sites/default/files/STRATEGIA_synteza%20ENG.pdf

Periodic Timetable Nonlinear Optimization in Public Transport Network

Marcin Kiciński

1 Introduction

Transport is a sector of the economy that integrates several sectors like manufacturing and services and thus contributes to its development. This integration is achieved because transport fulfils essential tasks, such as facilitating trade in goods, generating gross domestic product [1, 2], acting as a location factor, shaping spatial order. Transport is also vital in achieving the objectives relating to the functioning of the state and environmental protection [3, 4]. However, one of its main tasks involves meeting the communication needs of the population [5, 6]. In the case of urbanized areas, municipal transport is organized by local government units. Here, a unique role is played by public transport, in which passenger transport covers the area of the municipality/municipalities or agglomeration and serves the public according to a timetable (usually) defined by designated entities. The term 'public transport' means, in this case, that the service is carried out along predefined routes and according to a publicly available timetable, and that such service is available usually upon payment of a fare. Currently, the most common type of regular public transport in urban areas (cities and agglomerations) is the bus and tram transport.

Currently, the most common type of regular public transport in urban areas is the bus and tram transport. Both in the case of bus and tram transport, line and point infrastructures play an important role. As regards point infrastructure, is an element which significantly limits the development of transport, as it requires relatively high investment, both in terms of construction and maintenance [5–7]. Each of the aforementioned elements has its given capacity defined as the maximum number of means of transport which may pass through a particular transport element of linear or point

M. Kiciński (✉)
Poznan University of Technology, Institute of Transport, Poznan, Poland
e-mail: marcin.kicinski@put.poznan.pl

G. Sierpiński and E. Macioszek (eds.), *Decision Support Methods in Modern Transportation Systems and Networks*, Lecture Notes in Networks and Systems 208,
https://doi.org/10.1007/978-3-030-71771-1_4

infrastructure in a unit of time (within an hour, a day or a year). Thus, with regard to tram transport, the following essential infrastructure elements can be distinguished:

- tracks,
- electric traction,
- tram loops,
- stops,
- depots.

From the point of view of the quality of service provided by public transport (bus, tram, railway), a number of factors need to be taken into account in its organization, for example: line/route planning [8–15], vehicle scheduling and composition [6, 16, 17], integration [7, 9, 17, 18], depots, stops, loops, interchange hubs locating [8, 19, 20], timetabling [5, 10, 11, 21–24] with synchronization. According to many researchers [6, 10, 17, 21, 22, 25], the correct synchronizations of line layouts shortens the waiting time. It allows for a more even distribution of passenger streams over vehicles, thus avoiding overloading some of them. The most significant inconveniences for passengers associated with incorrectly planned timetables include:

- overlapping courses of two lines,
- departure at the same time,
- long intervals between courses.

According to the authors, it is a challenging task to meet these requirements in a situation of a particularly limited tram network capacity.

2 The Definition of the Decision Problem

The problem of timetabling, discussed in this chapter, concerns the companies operating public collective tramway transport, who have at their disposal the following data concerning:

- The number of tramway lines.
- The structure of both point and line infrastructure network, i.e., the lengths of individual track sections, tram journey times of individual lines between stops, and the existing twisting relationships at the nodes.
- The same interval of tram traffic in the entire network.

Bearing in mind the above, the decision-making problem of determining timetables in the tramway public transport network consists of establishing the time at which the trams of individual lines will depart from their initial nodes. Thus, in certain conditions resulting from the existing network, it will be possible to synchronize the trams in individual intermediate nodes in the best possible way (here, the stops located at such nodes).

In the opinion of practitioners, proper synchronization involves an attempt to meet the previously mentioned passenger demands while maintaining the requirements of the public transport operator at a certain level. In their opinion, passengers prefer the lines with a large number of stops connecting the loops located on opposite sides of the city [25]. This model of tramway communication system makes it difficult to synchronize the lines as the delay of the departure from the loop by one minute causes implications at many nodes. Proper synchronization, therefore, depends on appropriate manipulation of the departure time of trams serving particular lines from the loop so that the following situations would not occur at the nodes:

- "Bumper to bumper rides"—that is the movement of trams of different lines in one minute through the same stop. This issue mainly concerns the larger junctions of the city, where trams can come from different directions, and the time to reach the said node is different which makes the trams arrive at the same node simultaneously. There is then a case of blocking the tram or, in extreme cases, blocking the intersections.
- "Duplication of courses"—an event in which trams serving different lines follow one another.

3 The Mathematical Model of the Decision Problem

3.1 Basic Information

The mathematical model of the decision problem discussed in this chapter was formulated as a single integer optimization task.

The input data for the mathematical model of tramway traffic synchronization include:

- The diagram of the tramway network, which illustrates the network of connections with the specified nodes and tram loops as well as the information on the lines that serve these nodes and loops.
- The record of the routes of individual lines, i.e. information on the operation of a given node by a line with a specified number.
- The information about the traffic interval used and the time of passage between successive.

For each node, data on the location of the stop is required along with the information on the defined direction of movement of trams. The model is marked as follows:

- Geographical direction indication—N1. The marking in relation to the windrose attached to the scheme signifies W (West), E (East), N (North), S (South).
- Marking of the direction of movement—N2. Adopted orientation: U (Up—north), D (Down—south), L (Left—west), R (Right—east), D (Down—south).

This way of marking allows specific information to be coded. For example—$N1$ $N2$—ND—the stop is located on the northern side of the node, tramway traffic takes place to the south.

3.2 Mathematic Formulation

Input data:

- I—a mean the traffic interval adopted in the diagram, where $I \in N_{-0}$,
- w—a node in the network, where $w = 1, 2, \ldots, W$,
- $p_{N1,N2,w}$—a specific stop at a w node with the position of the stop $N1$ and direction of traffic $N1$, numbers of stop in w node: $\sum_{N1}\sum_{N2}\sum_{w} p_{N1,N2,w}$,
- $Z_{p_{N1,N2,w}}$—a set of tram lines serving a stop $p_{N1,N2,w}$,
- $x_{n,p_{N1,N2,w}}$—a minute of departure of n particular line from a stop $p_{N1,N2,w}$,
- n—a number of tram line, where $n = 1, 2, \ldots, N$
- pe—a tram loop, where $pe = 1, 2, \ldots, PE$,
- $t_{pe,w}$—a time of the journey from the tram loop pe to the node w,
- $t_{w,w+1}$—a time of the journey from the node w to the node $w+1$,
- $O_{p_{N1,N2,w}}$—a set of all departures from the stop $p_{N1,N2,w}$,
- $K_{Z_{p_{N1,N2,w}}}$—a set of indicators of natural numbers assigned to the individual elements of the $Z_{p_{N1,N2,w}}$ and their corresponding presence, where $K_{Z_{p_{N1,N2,w}}} \in \{k_1, k_2, \ldots, k_N\}$.

The decision variables for this model are the values of the departures of the individual trainsets from the loop (starting nodes—pe)—$y_{n,pe}$, where $y_{n,pe} \in N$. It should be remembered that even the accuracy of 1 min is often difficult to achieve in the case of trams due to, for example, changes in passenger boarding and alighting times.

In the mathematical model, the following set of constraints was included:

- The decision variables cannot exceed its the traffic interval adopted in the network:

$$y_{n,pe} \leq I \tag{1}$$

The synchronization is created for one interval, and it implies cyclical repeatability of courses within the time specified therein.

- The departure time of trams from each loops are different (in the interval I):

$$\bigwedge_{pe=a} y_{n,pe} \neq y_{n',pe} \tag{2}$$

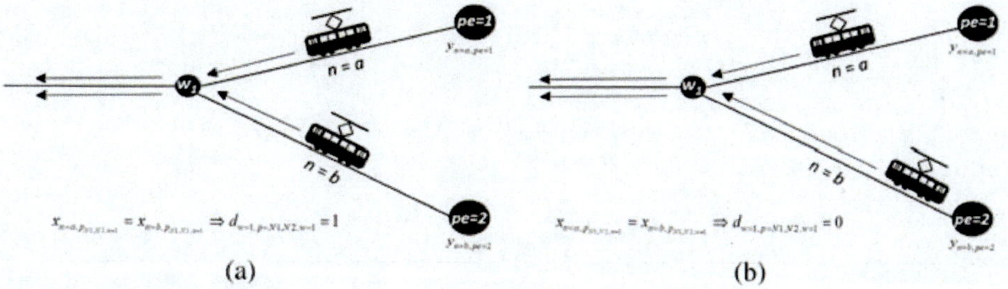

Fig. 1 Hypothetical situations at the tram stop: **a** $d_{w,p} = 1$, **b** $d_{w,p} = 0$

In addition, it is possible to include an optional limitation—providing a minimum of 3 min delay between the arrival on the loop and the departure of the same tram from the loop. This time is needed by the driver for maintenance activities, such as inspecting the interior of the tram.

In the proposed timetable synchronization model, one optimization criterion was suggested—minimizing duplication of journeys, i.e., as mentioned earlier, the passage of trams of different lines in one minute via the same stop.

$$\min \sum_{w} \sum_{p} d_{w,p} \tag{3}$$

where:

$$d_{w,p} = \begin{cases} 1, & \text{the movement of trams of different lines } w \\ & \text{in one minute through the same stop } p, \\ 0, & \text{otherwise.} \end{cases} \tag{4}$$

and

$$d_{w,p} = f(y_{n,pe}) \tag{5}$$

The idea of the $d_{w,p}$ presented in Fig. 1.

4 Case Study on the Example of the City of Poznań

4.1 Preparing the Model

The model was tested on the example of the tram network of the city of Poznań (Poland), whose diagram (constructed in MS Excel) is shown in Fig. 2.

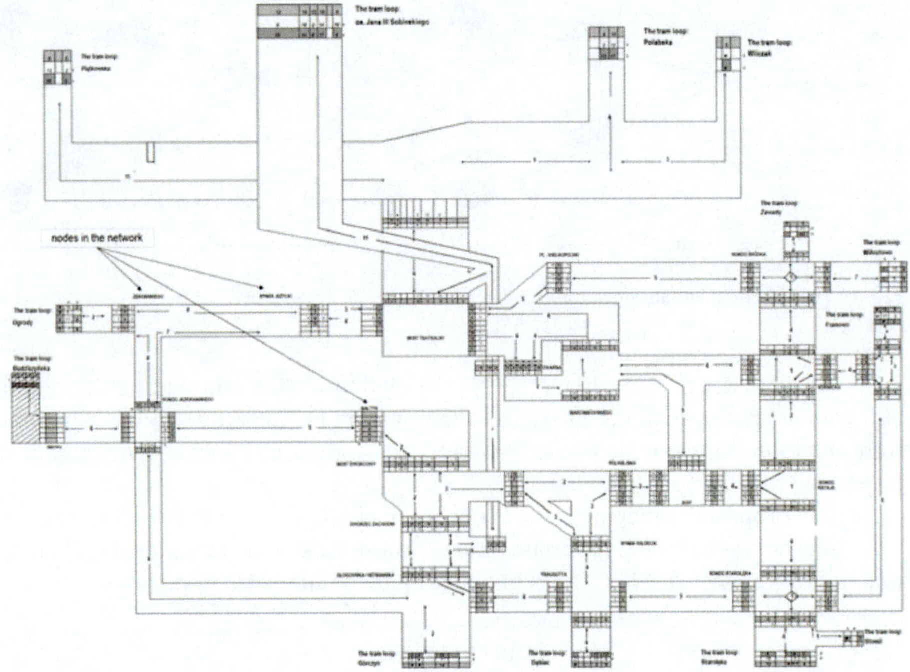

Fig. 2 Diagram of the discussed tram network (nodes and tram loops) of the city of Poznań (*source* MPK Poznań Sp. z o.o.)

The individual elements of the point and line infrastructure have been designed so that the results of the optimization procedure can be read directly. They are explained in Fig. 3.

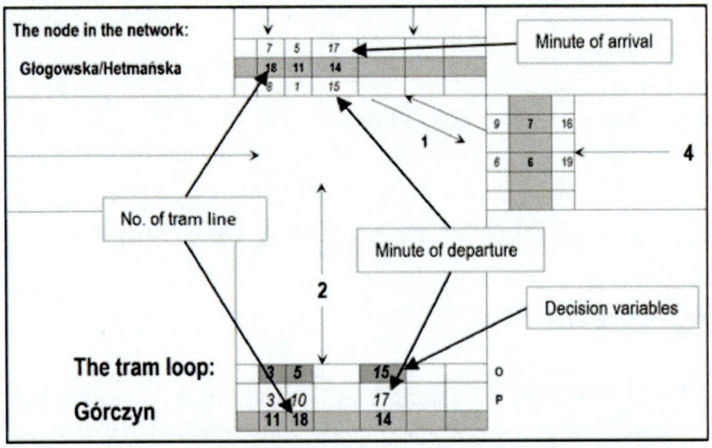

Fig. 3 Explanations of individual elements of tram infrastructure modelled in MS Excel

The designed diagram was the basis for the construction of the sheet, in which the optimization procedure was conducted directly. As can be seen in Table 1, it concerns a node called "Roundabout Starołęka" (Rondo Starołęka), where there are different directions (e.g. $N1 = \text{N}$ and $N2 = \text{U}$). Also, a section of the line numbering considered in the calculation experiment is shown.

The model for the city of Poznań assumes the presence of 12 decision-making variables ($|y_{n,pe} = 31|$). The assumed markings are presented in Table 2.

Table 1 Part of the prepared sheet designed to optimize the synchronization of traffic in the network—the node "Roundabout Starołęka"—minutes of arrivals tram from stops (nodes or tram loop)

N1	N2	No. tram line																
		2	3	4	5	6	7	8	9	10	11	12	13	14	15	16	17	18
N	U			5			6					11	5					
N	D			8			19					13	3					
W	L					17	0											
W	R					8	5											
E	L				6	16							0					
E	R				1	9							16					
S	U			4	0							10	4				15	
S	D			9	7							14	4				1	

Table 2 List of decision-making variables adopted for the Poznań city tram synchronization model

Name of tram loop	No. of tram loop—*pe*	No. of tram lines departing from the loop	Decision-making variables: minutes of departure from a specific tramway loop
Piątkowska	1	5	$y_{1,1}$
		9	$y_{2,1}$
Dębiec	2	2	$y_{1,2}$
		9	$y_{2,2}$
		10	$y_{3,2}$
Starołęka	3	4	$y_{1,3}$
		12	$y_{2,3}$
		17	$y_{3,3}$
…	…	…	…
Wilczak	6	3	$y_{1,6}$
…	…	…	…
Górczyn	12	11	$y_{1,12}$
		14	$y_{2,12}$
		18	$y_{3,12}$

5 Preparing the Model

Computational experiments were carried out using the test version of the commercial solver—Evolver by Palisade Corporation version 7.6, in which optimization is carried out using evolutionary algorithms [26–28]. The Evolver is designed to solve problems that do not have a linear course [29].

The following parameters of the Evolver evolutionary algorithm were adopted in a computational experiment:

- crossing parameter—0.2/0.3/0.4,
- mutation parameter—0.1/0.2/0.3,
- population size—200, 500, 1000,
- optimization time 30 min,
- automatic speed.

In the calculation experiment, the starting point was assumed to include the initial values of variables from 1 to 15. The values signified simultaneous departures of all trams from the loop in 1st, 2nd, 3rd.... up to 15th minute of the assumed interval respectively (in this case, a twenty-minute interval). Apart from that, other limitations were applied so that it would be possible to search for the broadest possible range of acceptable solutions.

The optimization procedure yielded the minimum value of the criterion of course duplication. That value was 16. This means that for the adopted 15-minute interval it was not possible to find such a solution in which a disadvantageous situation when trams of different lines stopping in one minute in one place (at a stop) would not occur at any of the stops.

6 Conclusions

The synchronization of tram timetables is a very complex issue. It requires numerous factors to be taken into account, including those relating to the infrastructure. As the authors have shown, for the proposed model tested on the data of the tram network in Poznan, it is impossible, for a 15-minute interval, to plan the time of tram departure in such a way that, in theory, no adverse event occurs. The term 'adverse' understood as the same time of departure from the same stop. When taking into account the proposed model, it is possible to consider other criteria as well, e.g. minimizing the events in which trams of different lines follow one another. Also, from the point of view of the passenger, it is possible, according to the author, to propose such criteria, in which, e.g. five trams of different lines depart from a given stop in 1st, 4th, 7th, 10th and 13th minute for the whole network in the interval of 15 min for the whole network (minimizing) the spread of departure time between consecutive trams of different lines at the same stop in the assumed interval).

References

1. Jara-Diaz S (2007) Transport economic theory. Bingley, Emerald
2. Banister D (2002) Transport planning. Taylor & Francis, London
3. Sierpiński G, Macioszek E (2020) Equalising the levels of electromobility implementation in cities. In: Mikulski J (ed) Research and the future of telematics, communications in computer and information science, vol 1289. Springer, Heidelberg, pp 165–176
4. Macioszek E, Sierpiński G (2020) Charging stations for electric vehicles—Current situation in Poland. In: Mikulski J (ed) Research and the future of telematics, communications in computer and information science, vol 1289. Springer, Heidelberg, pp 124–137
5. Daganzo C, Ouyang Y (2019) Public Transportation systems: principles of system design. Operations planning and real-time control. World Scientific, New Jersey
6. Ceder A (2016) Public Transit planning and operation: modeling. Practice and behavior. CRC Press, Boca Raton
7. Kiciński M, Solecka K (2018) Application of MCDA/MCDM methods for an integrated urban public transportation system—Case Study. City of Cracow. Arch Transp 46:71–84
8. Schöbel A (2010) Optimization in public transportation: stop location. Delay management and tariff zone design in a public transportation network. Springer, New York
9. Ceder A, Wilson NHM (1986) Bus network design. Transp Res Part B: Methodol 20:331–344
10. Schiewe P (2020) Integrated optimization in public transport planning. Springer International Publishing, Cham
11. Schmidt ME (2014) Integrating routing decisions in public transportation problems. Springer, New York
12. Schmidt M, Schöbel A (2015) The complexity of integrating passenger routing decisions in public transportation models. Networks. 65:228–243
13. Canca D, De-Los-Santos A, Laporte G, Mesa JA (2019) Integrated railway rapid tran-it network design and line planning problem with maximum profit. Transp Res Part E: Logist Transp Rev 127:1–30
14. Larin O, Mavrin V, Almetova Z (2018) Simulation modeling for the evaluation of conflicts at stops of the urban route network. Transp Res Procedia 36:411–417
15. Vakulenko K, Kuhtin K, Afanasieva I, Galkin A (2019) Designing the optimal public bus routes network at suburban area. Transp Res Procedia 39:554–564
16. Davidich Y, Kush Y, Galkin A, Davidich N, Tkachenko I (2019) Improving of urban public transportation quality via operator schedule optimization. J Urban Environ Eng 13(1):23–33
17. Vuchic VR (2005) Urban transit: operations, planning and economics. Wiley, Hoboken, New York
18. Sawicki P, Kiciński M, Fierek S (2016) Selection of the most adequate trip-modelling tool for integrated transport planning system. Arch Transp 1:55–66
19. Sawicki P, Fierek S (2018) Mixed public transport lines construction and vehicle's depots location problems. In: Macioszek E, Sierpiński G (eds) Recent advances in traffic engineering for transport networks and systems. Springer International Publishing, Cham, pp 213–224
20. Canca D, Barrena E (2018) The integrated rolling stock circulation and depot location problem in railway rapid transit systems. Transp Res Part E: Logist Transp Rev 109:115–138
21. Dźwigoń W (2006) Time-Table synchronization in public transport. Transport Miejski i Regionalny 12:38–42
22. Jacyna M, Gołębiowski P (2015) An approach to optimizing the train timetable on a railway network. Presented at the Conference: Urban Transport 2015, València, Spain June 2
23. Ibarra-Rojas OJ, Giesen R, Rios-Solis YA (2014) An integrated approach for timetabling and vehicle scheduling problems to analyze the trade-off between level of service and operating costs of transit networks. Transp Res Part B: Methodol 70:35–46
24. Odijk MA (1996) A constraint generation algorithm for the construction of periodic railway timetables. Transp Res Part B: Methodol 30:455–464

25. Musialski K, Kiciński M (2015) Model optymalizacyjny synchronizacji linii tramwajowych. Celowość, efektywność i skuteczność projektu transportowego. Logika interwencji. SITK RP Oddział w Poznaniu, Poznań, pp 269–278
26. Bäck T, Fogel DB, Michalewicz Z (eds) Handbook of evolutionary computation. Institute of Physics Pub. Oxford University Press, Bristol; Philadelphia: New York
27. Michalewicz Z, Schmidt M (2003) Evolutionary algorithms and constrained optimization. In: Evolutionary optimization. Kluwer Academic Publishers, Boston, pp 57–86
28. Osyczka S, Krenich S (2006) Evolutionary algorithms for global optimization. In: Pintér JD (ed) Global optimization. Springer, US, pp 267–300
29. Simon D (2013) Evolutionary optimization algorithms: biologically-inspired and population-based approaches to computer intelligence. Wiley, Hoboken, New Jersey

Method of Evaluating Bus Stops Based on Safety Aspects

Agnieszka Tubis, Emilia Skupień, and Mateusz Rydlewski

1 Introduction

Bus stops are a key element of point infrastructure of the urban transport system. They are considered to have a major impact on increasing the attractiveness of public transport in terms of quality, accessibility and safety [1]. They are one of the basic elements integrating pedestrian traffic with public transport. For this reason, their location is of particular importance. Bus stops should be located in areas generating the greatest movement of people, i.e. mainly in the immediate vicinity of residential buildings. In line with the [2] guidelines, the bus stops should be located in points of housing complex that will serve the largest possible number of inhabitants. Another indicated area in which bus stops should be located are travel destinations (e.g. offices, shopping malls, universities, etc.). The areas where bus stops are located are therefore characterized by heavy traffic, involving a wide variety of participants—pedestrians, cyclists, public transport passengers, individual vehicles (passenger cars) and public transport vehicles. The occurring heavy traffic may increase the risk of adverse events in the bus stop zone. This is confirmed by the research presented in [3, 4]. Ulak et al. [3] found that accidents in the stop zone involving pedestrians were related not only to the traffic and road geometry, but also to the presence of facilities such as hospitals, supermarkets or religious buildings. By contrast, Zhao et al. [4] showed that public transport was significantly related to the location of educational buildings and shopping centres. According to Ulak et al. [5] the combination of the

A. Tubis (✉) · E. Skupień · M. Rydlewski
Wroclaw University of Science and Technology, Wroclaw, Poland
e-mail: agnieszka.tubis@pwr.edu.pl

E. Skupień
e-mail: emilia.skupien@pwr.edu.pl

M. Rydlewski
e-mail: mateusz.rydlewski@pwr.edu.pl

G. Sierpiński and E. Macioszek (eds.), *Decision Support Methods in Modern Transportation Systems and Networks*, Lecture Notes in Networks and Systems 208,
https://doi.org/10.1007/978-3-030-71771-1_5

results from both of these studies indicates that pedestrians around bus stops might be exposed to an elevated crash risk. The relationship between the location of a bus stop and accidents involving pedestrians was also confirmed in the research by Eom et al. [6]. Authors used this relationship to develop a model in order to optimize the bus stop locations on a given roadway.

For this reason, a key aspect of the operation of bus stops is to ensure an appropriate level of safety for various participants of the transport system and in the vicinity of the bus stop. Bus stops must therefore be carefully designed to ensure maximum safety for all traffic users, especially the weakest [1]. The layout of the bus stop should enable safe and smooth flow of both buses and passengers in the stop zone. Passengers must be able to board and leave the vehicle efficiently without significant difficulties or delays in adjacent traffic [7]. The safety of bus stops is of key importance not only for passengers and carriers, but also for the local community and the manager of the bus stop zone [8].

Indicated by Tiboni et al. [1] the need to protect the weakest concerns in particular people with reduced mobility and children. For this reason, many publications that have appeared, in particular in recent years, concern the safe use of people with reduced mobility from collective transport services. Examples of such research are described in [9–13]. More environmental friendly solutions have impact for the case [14, 15]. The second group of travellers to whom numerous studies described in the literature relate are children. In particular, the risks related to their safe reaching the bus stop are analysed. The research that described this issue was presented, among others, in [16–18].

In the literature, one can find numerous publications devoted to the assessment of bus stops in relation to individual stops [19–21], as well as bus stops constituting an element of interchanges node [22, 23]. In most of these publications, safety appears as one of the many criteria for the assessment. However, the research described in [19] deserves attention. Based on the risk concept, the authors developed an algorithm for determining the *Level of safety*. The determination of this level was based on three parameters: probability of occurrence of harmful events, intensity of the damage, that may result and safety index, which relates geometric features with car accidents. Interesting research in this area is also presented in [24]. The authors proposed in this article the concept of level-of-safety for bus stops. It is introduced and corresponding models are proposed to quantify safety levels, which consider conflict points, traffic factors, geometric characteristics, traffic signs and markings, pavement conditions, and lighting conditions.

The research conducted by The Texas Transportation Institute [8] is also noteworthy. As part of the Transit Cooperative Research Program (TCRP), the research team developed guidelines for the location and design of bus stops in a variety of operating environments. The research team conducted a literature review, surveys and interviews with those identified as stakeholders, and a review of the manuals of 28 transit agencies with regard to bus stop location and design.

The conducted desk research allowed the development of guidelines for the safety assessment of bus stops. Therefore, the purpose of this chapter is to present the method of assessing bus stops in terms of the safety of their operation in the urban road transport system. This method is described in detail in Sect. 2 of the chapter. Then, Sect. 3 describes its use in the evaluation of selected bus stops in Wroclaw (Poland) and the results obtained from the measurement. The final conclusion and the summary are presented in Sect. 4.

2 Proposed Research Method

The aim of the chapter is to present the proposed method of assessing the safety of bus stops. The criteria for assessing safety were developed on the basis of a review of the literature on the safety of public transport stops (mainly bus stops, but also analyses of tram stops assessed) and guidelines appearing in the standards for assessing stops in selected cities (mainly Polish, but also in other countries).

The safety assessment criteria can be divided into three basic groups and 10 subgroups (Fig. 1):

A. The bus stop area.

- A.1 **Zone of the access to the bus stop** (A.1.1. the distance of a pedestrian crossing from the stop, A.1.2. Traffic lights at the pedestrian crossing, A.1.3. Clearly visible road signs for pedestrian crossings, A.1.4. Lighting of the walkway leading to the bus stop, A.1.5. Lighting of the road on which buses run, A.1.6. No stairs in the walkway to the bus stop, A.1.7. Presence of an asylum at a pedestrian crossing that separates the traffic lanes in opposite directions).
- A.2 **Presence of the other traffic users** (A.2.1. Permissible speed of vehicles in the vicinity of the bus stop, A.2.2. Close vicinity of the bus stop area).
- A.3 **Close surroundings** (A.3.1. The infrastructure around the bus stop obscures drivers from the stop, A.3.2. Advertisements in the area of the stop that distract drivers, A.3.3. Vehicles parked less than 10 m from the bus stop)

B. The passenger area.

- B.1 **The surface of the passenger waiting area** (B.1.1. Subsidence in the surface of the bus stop area, B.1.2. Dehydration in the bus stop area, B.1.3. Uniform surface of the stop)
- B.2 **Pedestrian traffic in rush hour** (B.2.1. Level of service in the area of the bus stop when changing passengers, B.2.2. Level of service on the pedestrian route adjacent to the stop at rush hour)
- B.3 **Adaptation to people with disabilities** (B.3.1. waiting area for people with reduced mobility, B.3.2. No stairs or downcast in the stop area, B.3.3. DIP board that can be read aloud at the bus stop, B.3.4. Protrusions surface at the edge of the platform).

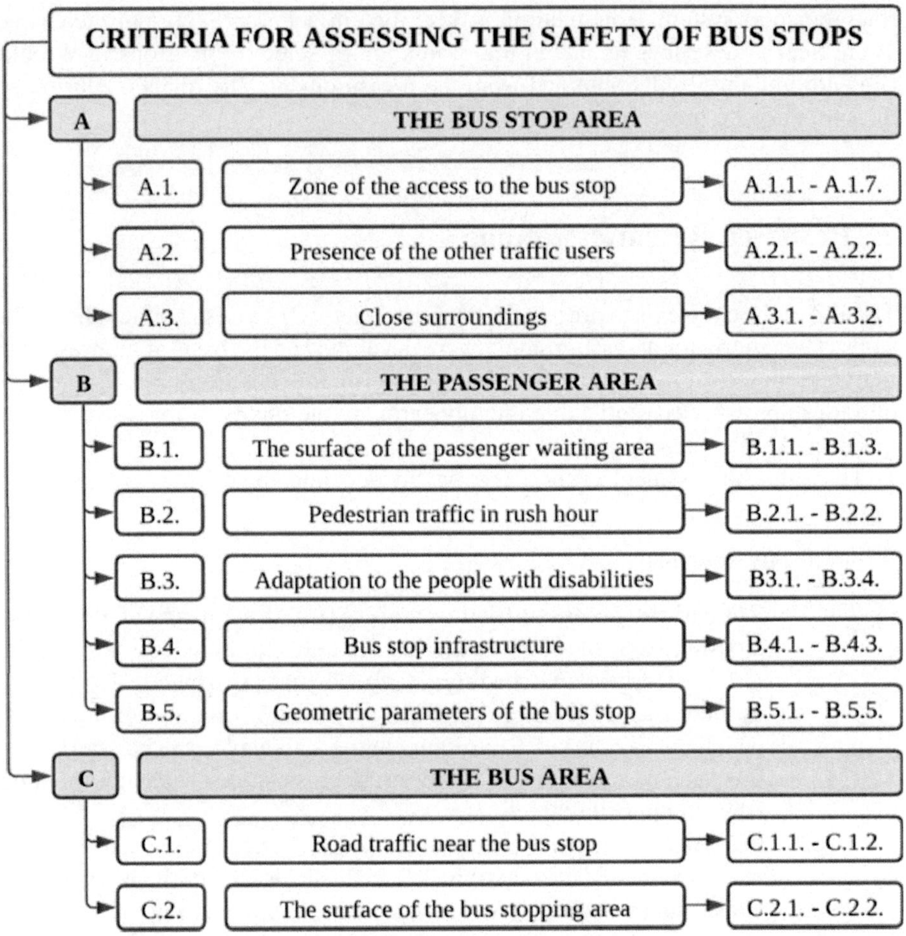

Fig. 1 Groups and subgroups of assessed criteria, *source* own study

B.4 **Bus stop infrastructure** (B.4.1. Lighting covering the entire bus stop area, B.4.2. Method of separating the area closely adjacent to the bus stop, B.4.3. Vertical sign informing about the bus stop)

B.5 **Geometric parameters of the bus stop** (B.5.1. Length of the stop, B.5.2. Width of the stop, B.5.3. Stop not on the curve of the road, B5.4. 2 m wide strip along the edge of the stop without any infrastructure obstacles, B.5.5. Curb elevation height relative to served rolling stock)

C. The bus area.

C.1 **Road traffic near the bus stop** (C.1.1. Number of lanes in- and in opposite direction of the bus route, C.1.2. Free possibility of joining the traffic by bus during rush hour)

C.2 **The surface of the bus stopping area** (C.2.1. Presence of cavities on the road in the bus stop area, C.2.2. Presence of ruts on the road in the bus stop area).

The assessment is carried out by the method of field research. This means that the person conducting the assessment should inspect the bus stop on the spot so as to be able to assess not only the condition of the bus stop infrastructure, but also make accompanying observations that will allow to assess the traffic characteristics (road, pedestrian and other moving participants) in the zone and around the bus stop. In order to facilitate the audit, a questionnaire form was developed for the proposed method. In addition to the assessment criteria discussed above, the form includes a form that allows to present the characteristics of the stop. In the form, the researcher enters:

- Name and number of the bus stop,
- Number of day and night lines served by the stop and the frequency of vehicle approach to the stop zone,
- Type of the bus stop
- Date and time of the questionnaire filling.

In the prepared questionnaire, most elements are assessed using the yes—no questions. Only the assessment of the intensity of passenger exchange uses the 5-point Level of Service scale. In two criteria concerning the identification of traffic participants and the method of separating the stop area, the researcher has an option of providing multiple answers, as well as using the "other" option. The 3 assessed criteria require the evaluator to provide a specific numerical value (number of lanes and permitted speed of vehicles in the vicinity). The adopted form of the assessment questionnaire allows the researcher to quickly register all measurements. However, relying on this requires the researcher to interpret the results at the stage of their analysis. For this reason, formulas have been prepared in the spreadsheet that assign the appropriate number of points to individual answers given in the points assessed.

For each of assessed bus stop, a safety index is calculated. It is the sum of the total points scored in all assessed criteria. On the basis of this indicator, the safety level of the stop is determined according to the scale presented in the Fig. 2.

On the basis of the defined safety level of the bus stop and the identified currently occurring deficiencies (limitations), decision-makers can initiate investments aimed at increasing the safety level.

0-40%	41-60%	61-80%	81-100%
not safe	requires improvements	safe but requires minor improvements	safe

Fig. 2 Safety rating scale of the bus stop according to the safety index, *source* own study

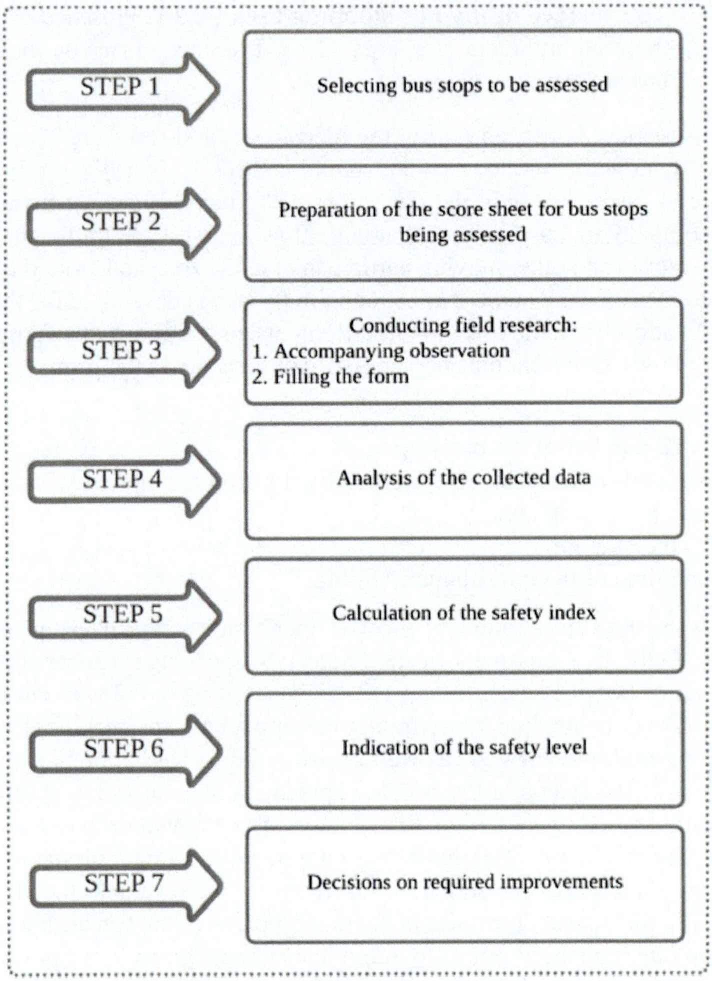

Fig. 3 Stages of the procedure for assessing the safety of bus stops, *source* own study

The stages of the research procedure in the proposed method of bus stop safety assessment are shown in Fig. 3.

3 Method Implementation at Selected Bus Stops—Evaluation Results

It was decided to present the proposed method of assessing the safety level of bus stops on the example of selected bus stops located in the southern part of Wroclaw, along Krzycka and Wałbrzyska streets. Two stops with passenger service in the bus

Table 1 Basic information about assessed bus stops, *source* own study

No.	Bus stop name	The place where the bus stops during the exchange of passengers	Number of daily lines served	Number of night lines served	Presence of a bus stop shelter
1	ZIMOWA	Bay	3	0	Yes
2	SKARBOWCÓW	Lane	3	1	No
3	KLECINA	Bay	3	1	Yes
4	KOŚCIELNA	Lane	3	1	No

bay and two with service directly from the street lane were selected for the analysis. Detailed information on the selected stops is presented in Table 1. Each of these stops is served by approximately 15 buses during rush hour. The audit based on the measurement form prepared for the method was carried out in October 2020.

The further part of the chapter presents the detailed results of the audits carried out, indicating the areas with irregularities observed at the analysed bus stops. Figures 4, 5, 6 and 7 presents the assessed bus stops.

Figure 4 shows the Zimowa bus stop. It is an example of a bus stop located at the stop bay in the vicinity of single-family houses. A pedestrian route runs through the stop area. The surface of the bus stop is made of uniform square flagstones. At the height of the approach slope of the stop bay, there is an entrance to a private position with a lowered pavement in this place. Street lighthouse are located on the

Fig. 4 "Zimowa" stop in Wrocław, *source* own study

Fig. 5 "Skarbowców" stop in Wrocław, *source* own study

opposite side of the road. Such a location of the lighthouse does not provide sufficient illumination of the entire road after dusk. The access area to the bus stop is provided by sidewalks along the road.

The next analysed bus stop is the "Skarbowców" stop shown in Fig. 5. Passenger service takes place when the bus stops on the road. The bus stop in the waiting area has only a bench without a roof. The bus stop is located in the vicinity of the square without adjacent buildings. As with the previous bus stop, there is a pedestrian crossing through the area. The surface of the stop has a non-uniform asphalt surface, which has numerous corrugations. Road lighting is located in close proximity to the stop.

Figure 6 shows the next analysed bus stop—Klecina. It is the second stop where passengers are exchanged directly on the road. The pedestrian route runs through the stop, and the bicycle traffic led in front of the stop is directed to the road, which can be seen in the downhill section shown in the lower left corner. The pavement surface is made of uniform paving stones. The lighting is located on the side of the bus stop.

The last of the analysed bus stops is the Kościelna stop—Fig. 7, which is located next to apartment blocks and service premises. Passenger exchange takes place at the bus bay. In the vicinity of the stop there is a wide strip that acts as a pavement. There are parked vehicles nearby and numerous service outlets. The road lighting is on the opposite side of the road..

Detailed information on the audit conducted at the analysed stops has been collected and presented in Table 2. The table contains detailed information on the implementation of the individual evaluation criteria used in the method. Points were

Fig. 6 "Klecina" stop in Wrocław, *source* own study

Fig. 7 "Kościelna" stop in Wrocław, *source* own study

Table 2 Detailed information about assessed bus stops, *source* own study

Assessment criterion	ZIMOWA	SKARBOWCÓW	KLECINA	KOŚCIELNA
A.1.7.	Lack	Lack	Presence (>2 m)	Presence (<2 m)
A.2.1.	30–50 km/h			
A.2.2.	Pavement	Walking cycle path	Pavement	Pavement
B.1.1.	Minor	Minor	Lack	Lack
B.2.1.	B or C			
B.2.2.	B or C			
B.4.2.	Lack	Lack	Lack	With a line
B.5.5.	Low	Lack	Low	Low
C.2.1.	Single	Single	Lack	Lack
C.2.2.	Single	Many	Lack	Lack
Criteria fully not met (individual)	A.1.4., B.1.2., B.3.3., B.4.1., B.5.2., C.1.2.	B.1.2., B.1.3., B.3.3., B.4.1., B.5.2.	B.3.3.	A.1.4., A.3.2., A.3.3., B.1.2., B.3.2., B.3.3., B.4.1.
Criteria fully not met (at all bus stops)	A.1.1., A.1.2., A.3.1., B.3.1., B.3.3., B.3.4., B.5.4.			
TOTAL POINTS:	23	22	33	26
PERCENTAGE RESULT:	49%	47%	70%	55%

awarded according to the rules set out in the method. The final score for each stop is at the bottom of the table.

Based on the analysis, based on 37 evaluation criteria, out of which it was possible to score 47 points, the best result among the analysed bus stops was the Klecina stop, which obtained 33 points (70%). Zimowa and Skarbowców stops are characterized by meeting the criteria below 50% of which the second stop fares worse. These results show a clear need for improvements to improve user safety. When analysing the obtained results, the main problem areas include:

- Maladjustment of the bus stop width,
- Lack of space free from infrastructure obstacles with a width of 2 m along the entire edge of the bus stop,
- No warning pavement strip along the edge of the bus stop platform,
- Raising the surface of the bus stop in relation to the road to an insufficient height,
- Commonality of the bus stop area with pedestrian routes (no pavement running outside the passenger waiting zone),
- Failure to provide adequate lighting for the bus stops, walkway to it and pedestrian crossings adjacent to the stop,
- The lack of drainage of bus stops and the roadway causing water to remain in the passenger exchange space and the possibility of splashing pedestrians,

- Failure to adapt the stops to serve people with special needs, e.g. no waiting zones, dynamic passenger information boards or the location of the bus stop shelter on an elevation (Kościelna stop),
- Presence of rut on the road in the bus stop area,
- The occurrence of single defects in the surface of the passenger exchange zone (most often large defects in the curbs at the edges of the platform).

Moreover, in close proximity of the analysed bus stops there are no pedestrian crossings. These crossings, apart from the fact that are located at a considerable distance from the stops, are not affected by traffic lights or, in some cases, are not equipped with an asylum between the lanes running in opposite directions. Lack of pedestrian crossings in the vicinity of stops may result in an increased number of illegal behaviours and crossing the road in a prohibited place.

4 Conclusions

The conducted research shows that individual bus stops even in one city have different safety standards. At the same time, the stops that have been assessed do not sufficiently meet the safety requirements that are imposed on public transport infrastructure points. On the basis of the proposed assessment method, all currently existing elements/conditions that threaten the safety of passengers or other traffic participants in the bus stop area have been clearly established. Thanks to this, the research results can be the basis for the development of possible improvements aimed at increasing safety in the area of assessed bus stops.

The proposed method, through the use of a formalized and simplified sheet to conduct an audit, is a simple tool that allows to assess the safety of bus stops. Thanks to the comprehensible and simple form of the questionnaire, the testers must have the basic knowledge required to collect key information on the measurement in the established groups for the assessment of stop safety aspects. The formalized system of given answers allows to go through the entire process of stop evaluation step by step. This solution also ensures that the auditor will pay attention to all the elements that, based on the literature research, have been considered important from the point of view of the safety of all traffic users within the bus stop. The adopted measurement model therefore reduces the burden (requirements) for the researcher collecting the data, and places greater emphasis on the requirements for the subsequent analysis carried out on the basis of the collected forms.

The calculated bus stop safety index can fulfil two roles. Firstly, thanks to the relation of its size to the proposed safety thresholds, this indicator can be the basis for benchmarking analyses carried out in relation to various bus stops. Secondly, it indicates the level of compliance with the safety criteria in relation to the designated safe bus stop pattern. Thanks to this, it is possible to indicate the necessary changes and improvements for the tested facilities, as well as the subsequent assessment of the undertaken investment activities in terms of their impact on the achieved safety level. It can be used to monitor the current state of stops, as well as provide a basis for planning future investments and designing new facilities. For this reason, the proposed assessment method can be use by owners and units managing point infrastructure in the city.

The presented method of assessing the safety of bus stops will be developed by the authors in two directions. First of all, in the next step, based on research among experts, weights to individual subgroups will be assigned. It will allow for the diversification of the importance of individual assessments. In the current method, the authors assumed that all assessed criteria are equal. As part of further work on improving the method, it can also be extended with additional evaluation elements, which, after grouping into specific thematic blocks, can provide a multifaceted assessment of the studied objects. Among such thematic groups, in addition to the safety presented in this chapter, one can focus on e.g. functionality, adapting the stop to the needs of people with reduced mobility or suffering from other types of disabilities, e.g. blind or deaf people.

References

1. Tiboni M, Rossetti S (2013) Implementing a road safety review approach for existing bus stops. WIT Trans Built Environ:699–709
2. Instytut Rozwoju Terytorialnego: Zintegrowany system powiązań miast w Legnicko-Głogowskim Obszarze Funkcjonalnym (LGOF) (2015) https://irt.wroc.pl/aktualnosci/download/file_id/310/doc_id/288.html. Accessed 30 June 2020
3. Ulak MB, Ozguven EE, Vanli OA, Dulebenets MA, Spainhour L (2018) Multivariate random parameter Tobit modeling of crashes involving aging drivers, passengers, bicyclists, and pedestrians: Spatiotemporal variations. Accid Anal Prev 121
4. Zhao J, Deng W, Song Y, Zhu Y (2013) What influences Metro station ridership in China? Insights from Nanjing. Cities 35:114–124
5. Ulak MB, Kocatepe A, Yazici A et al (2021) A stop safety index to address pedestrian safety around bus stops. Saf, Sci
6. Eom DL, Hong SW, Park SW (2014) The proper bus stop location from an intersection considering traffic safety. In: Proceedings of the transportation research board 93rd Annual Meeting
7. Karuppanan G, Prathibaa K, Gunasekaran K (2016) Planning of bus stops for safe and efficient passenger boarding and alighting. Int J Eng Res Technol 5(06):296–304
8. State of Florida Department of Transportation: Safer stops for vulnerable customers (2013) https://www.nctr.usf.edu/pdf/473-13.pdf. Accessed 30 June 2020
9. Peña Cepeda E, Galilea P, Raveau S (2018) How much do we value improvements on the accessibility to public transport for people with reduced mobility or disability? Res Transp Econ 69:445–452
10. Raczyńska-Buława E (2017) Dostosowanie usług transportu publicznego do potrzeb osób niepełnosprawnych. Autobusy 7–8:31–41
11. Solecka K, Nosal Hoy K, Deryło A (2020) Assessment of transport interchanges for the needs of people with reduced mobility. Travel Behav Soc 21:48–56
12. Naumov V (2018) Synchronisation of timetables for public bus lines using genetic algorithms and computer simulations. Lecture notes in networks and systems, vol 36. Springer, pp 44–53
13. Alrawi F, Ali A (2020) Analysis of the relationship between traffic accidents with human and physical factors in Iraq. In: Macioszek E, Sierpiński G (eds) Modern traffic engineering in the

system approach to the development of traffic networks. Advances in intelligent systems and computing, vol 1083. Springer, Cham. https://doi.org/10.1007/978-3-030-34069-8_15
14. Sierpiński G, Macioszek E (2020) Equalising the levels of electromobility implementation in cities. In: Mikulski J (ed) Research and the future of telematics, communications in computer and information science, vol 1289. Springer, Heidelberg, pp 165–176
15. Macioszek E, Sierpiński G (2020) Charging stations for electric vehicles—Current situation in Poland. In: Mikulski J (ed) Research and the future of telematics, communications in computer and information science, vol 1289. Springer, Heidelberg, pp 124–137
16. Unger R, Eder C, Mayr JM, Wernig J (2002) Child pedestrian injuries at tram and bus stops. Injury 33:485–488
17. Ampofo-Boateng K, Thomson JA (1991) Children's perception of safety and danger the road. Br J Psych 82:487–505
18. Sierpiński G, Staniek M (2016) Education by access to visual information – methodology of moulding behaviour based on international research project experiences. In: Gómez Chova L, López Martínez A, Candel Torres I (2016) ICERI2016 Proceedings, 9th International conference of education, research and innovation, 14–16 Nov 2016, pp 6724–6729. IATED Academy, Seville, Spain (2016)
19. Amadori M, Bonino T (2012) A methodology to define the level of safety of public transport bus stops, based on the concept of risk. Procedia Soc Behav Sci 48:653–662
20. Budzyński M, Tubis A, Rydlewski M (2020) Safety assessment method of public transport stops. Proceeding from 66. Konferencja Naukowa w Krynicy Zdrój
21. Koperski Ł, Żmuda-Trzebiatowski P (2018) Ocena stanu infrastruktury przystankowej. Studium przypadku: poznański łazarz. Pr. Nauk. Politech. Warsz Transp 121:181–190
22. Bryniarska Z, Zakowska L (2017) Multi-criteria evaluation of public transport interchanges. Transp Res Procedia. Elsevier B.V., pp 25–32
23. Dźwigoń W (2012) Kryterium bezpieczeństwa pasażerów w ocenie węzłów przesiadkowych. Transp Miej i Reg 06:13–18
24. Ye Z, Wang C, Yu Y et al (2016) Modeling level-of-safety for bus stops in China. Traffic Inj Prev 17:656–661

Methods of Transport Accessibility Management to Paid Parking Zones with Dynamic Parking Information

Agata Kurek

1 Introduction

Greater possibilities of owning a car change the behavior of road users, who more and more often use this means of transport when traveling. The increase in vehicles on city streets causes unfavorable road conditions, a decrease in road safety, mainly of unprotected road users, as well as a negative impact on the natural environment [1–7].

People managing traffic in the city are looking for solutions, the implementation of which may contribute to discouraging travel by a passenger car or to encourage the use of alternative means of transport. One of the solutions that can encourage drivers to travel to the city center by a means of transport other than a car are system parking such as P&R, B&R, etc. The purpose of these parking is to leave the car outside the most heavily trafficked area in the city and continue the journey using public transport, bicycle, on foot. Another tool is shared mobility, car-sharing, bike-sharing, scooter-sharing system, etc. Shared mobility allows you to rent a means of transport, e.g. a car, bicycle, scooter for a fee for a specified period [8–10].

The solutions that are to discourage drivers from traveling by car include the introduction of the Tempo 30 zone and the Woonerf zone. In the area covered by the first of these zones, there are physical limitations that force the driver to be careful and drive the vehicle at the appropriate speed [11, 12]. Woonerf, on the other hand, is a shared space zone that has the function of a street, parking lot, and a promenade [13]. In the case of this zone, there is no separation of road, pedestrian, and bicycle traffic. Other tools that can contribute to reducing traffic congestion in the city center

A. Kurek (✉)
Faculty of Transport and Aviation Engineering, Department of Transport Systems and Traffic Engineering, Silesian University of Technology, Silesian, Poland
e-mail: Agata.Kurek@polsl.pl

G. Sierpiński and E. Macioszek (eds.), *Decision Support Methods in Modern Transportation Systems and Networks*, Lecture Notes in Networks and Systems 208,
https://doi.org/10.1007/978-3-030-71771-1_6

are the introduction of parking charges, for entering the city center or a total ban on entering the city center [14, 15].

City authorities in Poland do not decide to close the entrance to the city center or to charge for entering the city. System parking, shared mobility and Paid Parking Zones (PPZ) can be found more and more often in cities. Paid Parking Zones are also being introduced to increase the availability of parking spaces in the city center for more drivers. However, a large number of cars cause traffic caused by searching for an empty place to leave the vehicle. The solution to this problem is the introduction of Dynamic Parking Information (DPI), the purpose of which is to inform drivers about empty parking spaces on variable message tables or via a mobile application.

The chapter aims to analyze the pedestrian accessibility of parking covered by the Paid Parking Zone and Dynamic Parking Information in the cities of the Górnośląsko-Zagłębiowska Metropolis (GZM). After the introduction, the second chapter presents the characteristics of PPZ and DPI operating in GZM cities. The third chapter presents an analysis of the pedestrian accessibility of parking covered by PPZ and DPI in GZM cities. The conclusions are presented at the end.

2 Transport Accessibility

According to [16], accessibility is the main product of the transport system, which determines the advantage of the location of a given area to other areas. It can be divided into several types:

- direct and indirect,
- relative and total,
- physical, temporary and economic,
- active and passive.

The methods of research and measurement of transport accessibility can be divided into [17]:

- availability measured by infrastructure equipment,
- potential availability,
- availability measured by distance,
- availability measured by isochrones,
- availability measured in the geography of time or space-time,
- availability measured by utility maximization.

The chapter presents the results of the pedestrian accessibility analysis measured with isochrones. It is the accessibility measured by the range of equal communication impact and in many cases may be a variant of the accessibility measured by distance, because, from the cartographic point of view, the isochrones method consists in designating zones with the same time distance.

No possibility to build new infrastructure due to the existing buildings and infrastructure of underground connections and the limitation of the use of a car when traveling, which is an element of high absorption of the area compared to the number of people transported, makes it necessary to distinguish zones in which the accessibility to the infrastructure of individual forms of transport is determined [18]:

- zone "0" (strict city center)—corresponding to the greatest accessibility to pedestrians;
- zone "A" (downtown area)—with limited access to cars, with large use of public transport;
- zone "B" (housing estates, industrial areas)—access to public and individual transport at the same level;
- zone "C" (areas with extensive development)—the greatest accessibility of car.

A characteristic feature of this division is the reduction of access to infrastructure for individual transport as one approaches the center while increasing the share of public transport. On the other hand, as moving away towards the "sleeping" areas of the city, the possibility of using individual transport increases, and the demand for public transport decreases. In the recent years, along with the development of electromobility and electric vehicles issues, scientific literature includes research works related to the analysis of the accessibility to charging points for electric cars from specific locations, e.g. places of residence, work, commercial and service facilities [19, 20].

3 Characteristics of Parking in the Conditions of PPZ and DPI Functioning in GZM Cities

The Górnośląsko-Zagłębiowska Metropolis (GZM) comprises 41 communal located in the Silesian Voivodeship in Poland. There is a Paid Parking Zone in 6 cities (Bytom, Chorzów, Gliwice, Katowice, Tychy, Zabrze), while 3 of them are equipped with systems that inform drivers about empty parking spaces—DPI (Chorzów, Gliwice, Piekary Śląskie). Table 1 presents cities of GZM with a Paid Parking Zone or Dynamic Parking Information and their characteristics.

The areas where the PPZ operates are characterized by a shortage of parking spaces. Parking fees in PPZ in all cities are collected on working days, from the morning hours (depending on the city: 09:00 or 10:00) to the afternoon hours (depending on the city: 17:00 or 18:00). The city of Gliwice has the largest number of streets covered by the PPZ, and Zabrze—the smallest number of streets covered by the PPZ. In Chorzów, Gliwice, and Katowice, PPZ is divided into subzones, each of them characterized by different parking fee rates. The highest parking fees are in subzone A in Gliwice and Katowice. Whereas the smallest in Bytom and Zabrze.

Figure 1 shows the area of the Metropolis GZM with the cities marked with PPZ and DPI. It can be noticed that the cities where parking in the city center are covered

Table 1 Characteristics of PPZ and DPI in GZM cities, *source* [21–26] (access: 29.12.2020)

City	Area	Parking fee	Fee collection hours	DPI occurrence
Bytom	39 streets	1,50 PLN—fee for the first hour of parking, 1,80 PLN—fee for the second hour of parking, 2,10 PLN—fee for the third hour of parking, 1,50 PLN—fee for the fourth and subsequent hour of parking.	On working days from 09:00 to 17:00	–
Chorzów	Subzone A:20 streets Subzone B: 12 streets	SUBZONE A 0,50 PLN—fee for the minimum parking time—15 min, 2,00 PLN—fee for the first, fourth, and each subsequent hour of parking, 2,20 PLN—fee for the second hour of parking, 2,40 PLN—fee for the third hour of parking. SUBZONE B 0,30 PLN——fee for the minimum parking time—15 min, 1,20 PLN—fee for the first, fourth, and each subsequent hour of parking, 1,40 PLN—fee for the second hour of parking, 1,60 PLN—fee for the third hour of parking.	On working days from 10:00 to 18:00	Number of empty parking spaces displayed on the website

(continued)

Table 1 (continued)

City	Area	Parking fee	Fee collection hours	DPI occurrence
Gliwice	Subzone A: 29 streets Subzone B: 57 streets	SUBZONE A: 0,70 PLN—fee for the minimum parking time—15 min, 1,40 PLN—fee for the first 0.50 h of parking, 2,80 PLN—fee for the first hour of parking, 6,10 PLN—fee for the second hour of parking, 10,00 PLN—fee for the third hour of parking, 12,80 PLN—fee for the fourth hour of parking. SUBZONE B: 0,40 PLN—fee for the minimum parking time—15 min, 0,80 PLN—fee for the first 0.50 h of parking, 1,60 PLN—fee for the first hour of parking, 3,50 PLN—fee for the second hour of parking, 5,70 PLN—fee for the third hour of parking, 7,30 PLN—fee for the fourth hour of parking	On working days from 09:00 to 17:00	Number of empty parking spaces on the website, in the application, and on the VMS

(continued)

by PPZ and DPI are located in the central part of the GZM area. Only the city of Tychy is distant from these cities and is situated near the southern border of the area.

Table 1 (continued)

City	Area	Parking fee	Fee collection hours	DPI occurrence
Katowice	Subzone A: 20 streets Subzone B: 2 streets	SUBZONE „Śródmieście" 2,00 PLN—fee for the first hour of parking, 2,40 PLN—fee for the second hour of parking, 2,80 PLN—fee for the third hour of parking, 2 PLN—fee for each subsequent hour of parking. SUBZONE „Ceglana—Wita Stwosza". 3,00 PLN—fee for the first hour of parking, 3,60 PLN—fee for the second hour of parking, 4,20 PLN—fee for the third hour of parking 3 PLN—fee for each subsequent hour of parking,	On working days from 09:00 to 16:30	–
Piekary Śląskie	–			2 VMS showing the number of empty places on 2 streets
Tychy	8 zones	1,00 PLN—fee for the first 0.50 h of parking, 2 PLN—fee for the first hour of parking, 2,40 PLN—fee for the second hour of parking, 2,80 PLN—fee for the third hour of parking, 2,00 PLN—fee for each subsequent hour of parking.	On working days from 09:00 to 17:00	–

(continued)

Table 1 (continued)

City	Area	Parking fee	Fee collection hours	DPI occurrence
Zabrze	2 streets and place	1,50 PLN—fee for the first hour of parking, 1,80 PLN—fee for the second hour of parking, 2,00 PLN—fee for the third hour of parking, 1,50 PLN—fee for the fourth and subsequent hour of parking.	On working days from 08:00 to 18:00	–

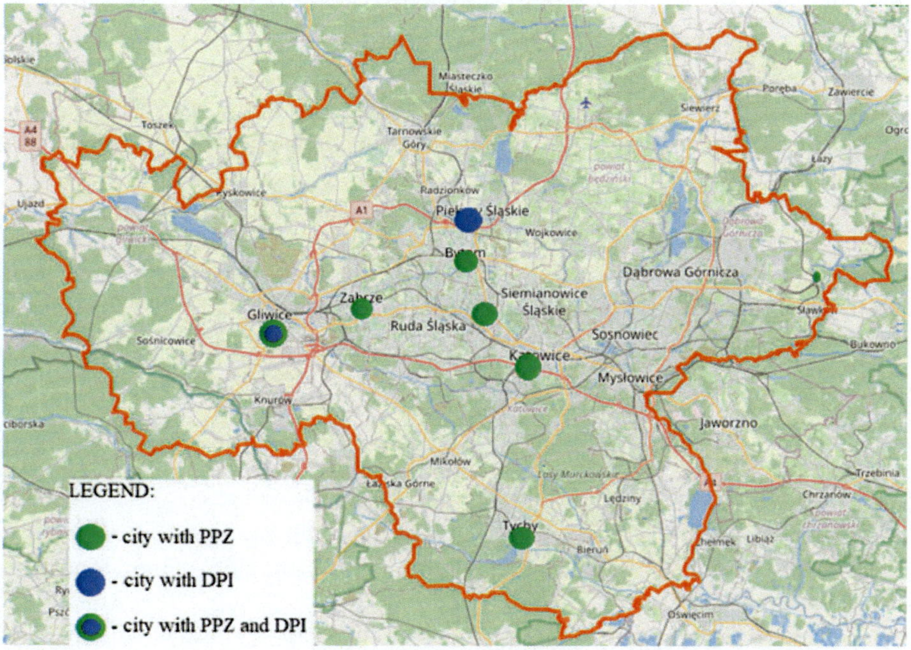

Fig. 1 GZM area and cities in which PPZ and DPI functioning, source: own research based on [27]

4 Analysis of the Pedestrian Accessibility of Parking in the Conditions of DPI Functioning in Gliwice and Piekary Śląskie

This chapter presents maps of the pedestrian accessibility of parking covered by the Dynamic Parking Information: Gliwice and Piekary Śląskie. In Chorzów, variable message tables indicate empty parking spaces that are located outside the road lane. Pedestrian accessibility has been determined in 3 pedestrian access variants: 5, 10, and 15 min. Pedestrian access isochrones were determined using the QGIS program. Designated isochrones take into account the existing buildings and linear transport infrastructure. Pedestrian accessibility maps allow for an analysis of the spatial development of nearby parking. Thanks to this, it is possible to conclude the destinations of the people using the analyzed parking.

There are 18 sections of streets in Gliwice where the DPI operates. On the other hand, in Piekary Śląskie, the DPI system indicates empty parking spaces on 3 street sections.

In 5 min walk from the analyzed parking in Gliwice, there are the market, commercial and service buildings, hotels, and the City Office. In the case of isochrones of 10 min of walking accessibility, they also include the railway station and the campus of the Silesian University of Technology. However, in the case of a 15-minute walk, the hospital is also available.

On the other hand, when analyzing the pedestrian accessibility map for the analyzed parking in Piekary Śląskie, there are commercial and service buildings, a hospital, a hotel, and the City Office within a 5-minute walk. However, within 10 min walking distance there are allotment gardens, a cemetery, and a supermarket. The 15-minute walking option also includes a church and a park (Figs. 2 and 3).

5 Conclusions

The chapter aimed to analyze the pedestrian accessibility of parking covered by the Paid Parking Zone and Dynamic Parking Information in the cities of the Górnośląsko-Zagłebiowska Metropolis.

The characteristics of parking presented in the chapter in the conditions of the functioning of PPZ and DPI in GZM cities and the analysis of the pedestrian accessibility of parking in the conditions of operation of DPI allowed for the formulation of the following conclusions:

- out of 41 communal included in the GZM, only 6 have PPZ, and 3 of them—DPI,
- PPZ operating in GZM cities are characterized by similar parking fee collection times, while the rates vary between cities,
- cities, where PPZ and DPI function, are located mainly in the central part of the GZM area,

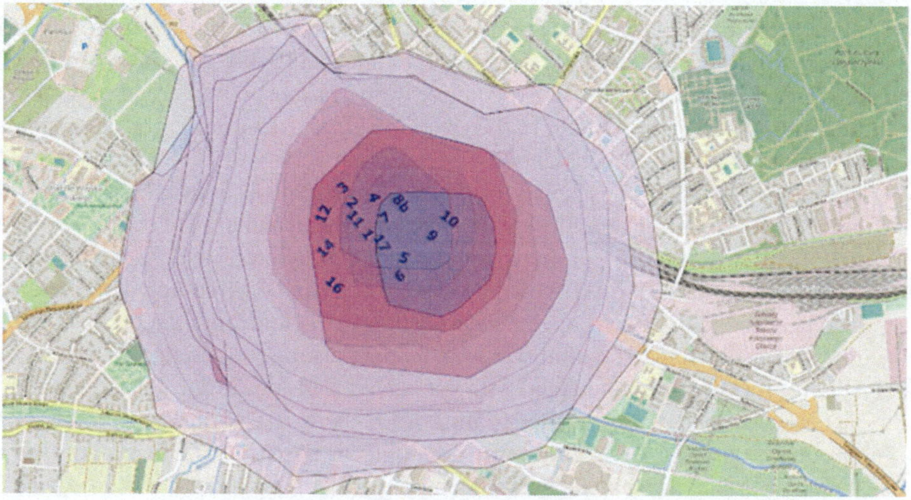

Fig. 2 Pedestrian accessibility of parking covered by the DPI in Gliwice

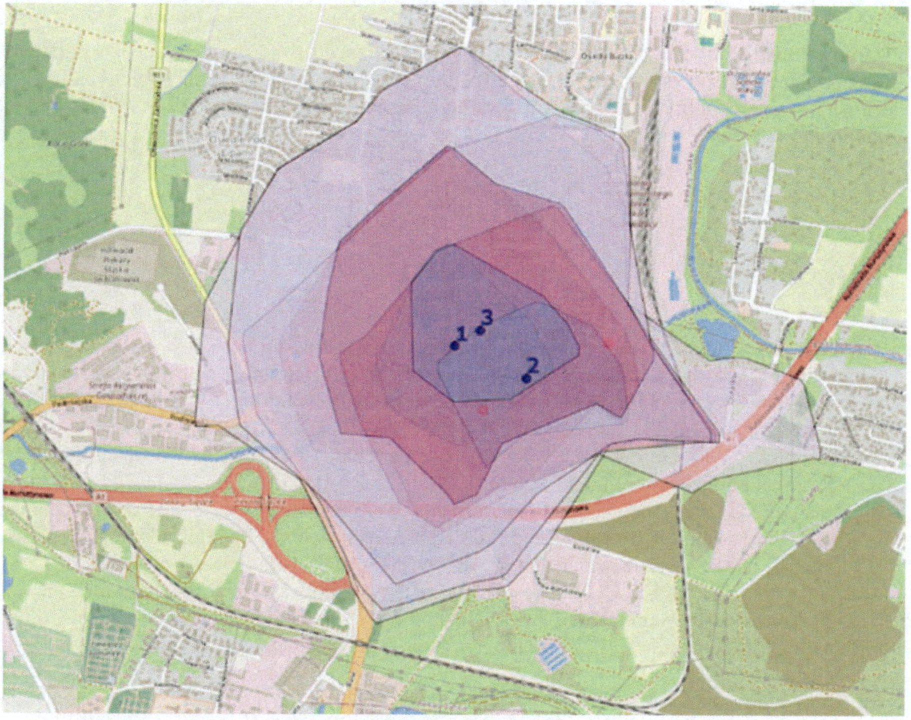

Fig. 3 Pedestrian accessibility of parking covered by the DPI in Piekary Śląskie

- the analysis of isochrones in variants 5, 10, 15 min of pedestrian access for parking covered by the PPZ and DPI in Gliwice indicates that the destination of people using parking may be mainly (5-minute walk variant) commercial and service facilities, hotels, City Office. In the case of a 10 and 15-minute walk, it can also be a railway station, a campus of the Silesian University of Technology, hospital,
- the analysis of isochrones in variants 5, 10, 15 min of pedestrian access for parking covered by DPI in Piekary Śląskie indicates that the destination of people using the parking may be mainly (5-minute walk variant) commercial and service facilities, a hotel a hospital, City Office. In the case of a 10 and 15-minute walk, these can also be allotment gardens, a cemetery, a supermarket, a church, a park,
- maps of pedestrian accessibility for analyzed parking indicte that in the case of Gliwice they may be associated with short-term parking. At the analyzed parking in Gliwice, apart from the functioning DPI, there is also the PPZ. Which can discourage drivers from long-term parking. On the other hand, in the case of Piekary Śląskie, the spatial development (e.g. allotment gardens, a church, a park) nearby may indicate longer parking. There is no PPZ in the analyzed parking in Piekary Ślaskie.

Acknowledgements The present research has been financed from the "Excellence Initiative—Research University" program implemented at the Silesian University of Technology,.2020-2022 as a part of a grant entitled "Analysis of parking characteristics in the conditions of PPZ and DPI functioning in selected areas of GZM cities".

References

1. Macioszek E (2018) The comparison of models for follow-up headway at roundabouts. In: Macioszek E, Sierpiński G (eds) Recent advances in traffic engineering for transport networks and systems. lecture notes in networks and systems, vol 21. Springer International Publishing Switzerland, pp 16–26
2. Szczuraszek T, Macioszek E (2013) Proportion of vehicles moving freely depending on traffic volume and proportion of trucks and buses. Baltic J Road Bridge Eng 8:133–141
3. Macioszek E (2012) Geometrical determinants of car equivalents for heavy vehicles crossing circular intersections. In: Mikulski J (ed) Transport Systems telematics. Communications in computer and information science, vol 329. Springer, Berlin Heidelberg, pp 221–228
4. Macioszek E, Sierpiński G, Czapkowski L (2010) Problems and issues with running the cycle traffic through the roundabouts. In: Mikulski J (ed) Transport Systems telematics. Communications in computer and information science, vol 104. Selected Papers. Springer, Berlin Heidelberg, pp 107–114
5. Kurek A, Juzyniec J, Kielc B (2018) Analysis of road safety in the city of Sosnowiec in period 2006-2016, Sci J Silesian Univ Technol Ser Transp 100:69–79
6. Wiedemann R (2019) Evaluation of the improvement of organisation and traffic control system on selected Part of the Transport Network. Współczesne problemy transportu: opracowanie monograficzne. T. 5, Sterowanie ruchem. Koło Naukowe Systemów Komunikacyjnych Politechniki Krakowskiej im. T. Kościuszki, pp 20–33 (in Polish)
7. Pashkevich A, Bairamov E, Burghardt TE, Sucha M (2020) Finding the way at Kraków Główny Railway Station: preliminary eye tracker experiment. In: Sierpiński G (ed) Smart and green

solutions for transport systems. Advances in intelligent systems and computing, vol 1091. Springer, Cham, pp 238–253. https://doi.org/10.1007/978-3-030-35543-2_19
8. Turoń K, Sierpiński G, Tóth J (2019) Support for pro-ecological solutions in smart cities with the use of travel databases–a case study based on a bike-sharing system in Budapest. In: Scientific and technical conference transport systems theory and practice, vol 1091, pp 225–237. Springer, Cham
9. Shaheen S, Cohen A, Chan NA (2020) Bansal. Sharing strategies: carsharing, shared micromobility (bikesharing and scooter sharing), transportation network companies, microtransit, and other innovative mobility modes, Transportation, land use, and environmental planning, pp 237–262
10. Clayton W, Ben-Elia E, Parkhurst G, Ricci M (2014) Where to park? A behavioural comparison of bus Park and Ride and city centre car park usage in Bath, UK. J Transp Geogr 36:124–133
11. Villarreal Chavez DB, Kurek A, Sierpiński G, Jużyniec J, Kielc B (2016) Review and comparison of traffic calming solutions—A case study for Mexico City and Katowice. Sci J Silesian Univ Technol Ser Transp 93:11–21
12. Sołowczuk A (2019) Effect of traffic calming measures implemented on the approach to the Tempo–30 zone on the Degree of speed reduction. IOP conference series: materials science and engineering 603:1–9
13. Dudek J (2009) Design guidelines for creating a vital woonerf street. Sofia: In: International multidisciplinary scientific Geoconference: SGEM. surveying geology & mining ecology management, vol 19, pp 433–441
14. Litman T (2010) Parking pricing implementation guidelines. Victoria transport policy institute
15. Nieuwenhuijsen M, Bastiaanssen J, Sersli S, Waygood EOD, Khreis H (2019) Implementing car-free cities: rationale, requirements, barriers and facilitators. In: Nieuwenhuijsen M, Khreis H (eds) Integrating human health into urban and transport planning. Springer, Cham, pp 199–219
16. Spiekermann K, Neubauer J (2002) European accessibility and peripherality: concepts, models and indicators. Nordregio, Sweden
17. Komornicki T, Śleszyński P, Rosik P, Pomianowski W (2009) Dostępność przestrzenna jako przesłanka kształtowania polskiej polityki transportowej. Biuletyn Komitetu Przestrzennego Zagospodarowania Kraju PAN 241:6–163
18. Szołtysek J (2007) Podstawy logistyki miejskiej. Katowice, Wydawnictwo Akademii Ekonomicznej w Katowicach
19. Sierpiński G, Macioszek E (2020) Equalising the levels of electromobility implementation in cities. In: Mikulski J (ed) Research and the future of telematics, vol 1289. Commun Comput Inf Sci. Springer, Heidelberg, pp 165–176
20. Macioszek E, Sierpiński G (2020) Charging stations for electric vehicles—Current situation in Poland. In: Mikulski J (ed) Research and the future of telematics, vol 1289. Communications in computer and information science. Springer, Heidelberg, pp 124–137
21. Municipal Road and Bridge Administration in Bytom: www.mzdim.bytom.pl
22. Municipal Streets and Bridge Administration in Chorzów: www.mzuim.chorzow.eu
23. Parking Gliwice: www.parkowaniegliwice.pl
24. Municipal Streets and Bridge Administration in Katowice: www.mzum.katowice.pl
25. Municipal Streets and Bridge Administration in Tychy: www.mzuim.tychy.pl
26. Municipal Roads and IT Infrastructure Administration: www.mzdii.zabrze.pl
27. Open Street Map: www.openstreetmap.org

Method to Evaluate a Bike-Sharing System Based on Performance Parameters

Anton Pashkevich, Marcin J. Kłos, Rafał Jaremski, and Meruyert Aristombayeva

1 Introduction

The intensive development of cities affects the mobility of residents. Large urban centres have many destinations for travel. The growing number of cars have negative impact on the environment (e.g. air pollution, noise) [1]. A challenge for city leaders is to reduce that phenomenon. There are many ways to change passengers' travel habits, e.g., pricing for parking, car-free and traffic-calming zones in city centre, advertising campaigns to promote different modes of transport. An important aspect is the sustainability of the transport means used (e.g. [2–5]). The development of the city and public transport should be planned in such a way that it is functional for users. For example, high accessibility to bus stops, low prices and high quality could bring new passengers to the public transport.

To address the above-mentioned challenge, it is necessary to promote a bike as a mode of transport, which is clean type of transport (without pollution). This kind of transport could also reduce traffic flow in the city. Users can get to their destination more accurately (e.g. bike parking in front door of office) and there is no problem with parking places [6]. To promote this mode of transport, many cities start to build

A. Pashkevich (✉) · R. Jaremski
Politechnika Krakowska, ul. Warszawska 24, 31-155 Kraków, Poland
e-mail: apashkevich@pk.edu.pl

R. Jaremski
e-mail: rafal.jaremski@gmail.com

M. J. Kłos
Politechnika Śląska, ul. Krasińskiego 8, 40-019 Katowice, Poland
e-mail: marcin.j.klos@polsl.pl

M. Aristombayeva
Kazakh-British Technical University, Tole Bi Str. 59, 050000 Almaty, Kazakhstan
e-mail: mika.aristombaeva@gmail.com

G. Sierpiński and E. Macioszek (eds.), *Decision Support Methods in Modern Transportation Systems and Networks*, Lecture Notes in Networks and Systems 208,
https://doi.org/10.1007/978-3-030-71771-1_7

own bike-sharing system. Governments of the cities must also remember to build additional infrastructure: bicycle paths and, e.g., bike-and-ride stations in places where it is possible to change means of transport [7]. Currently, there are many different systems in terms of operation and payment methods [8]. The city when deciding to install a bike-sharing system must pay attention on specific points where bicycle stations could have the limited accessibility for users. To solve that problem, it is important to investigate better locations for infrastructure, e.g., with using GIS data [9, 10]. The positive aspect of such station-based bike-sharing systems is the lack of bicycles left in the city everywhere: e.g., residents do not have to skip bikes left on the sidewalks. The other positive feature of docked bike-sharing system is the ease of maintenance by checking every dock (station) [11]. However, a dockless bike-sharing system has greater accessibility parameters for users. In combination with the application, it is easy to find the free bike close by [12]. On the other hand, in such a system it is harder to maintenance bikes, because an operator needs to reach all of them in different locations. Payment methods could be differing: e.g., special application for smartphone, subscription system or connection with public transport payment system.

In the chapter authors presented the method to evaluate bike-sharing systems based on performance parameters. The aim of the proposed method is to indicate basic parameters, which can help to assess the existing bike-sharing system from the perspective of a service provider as well as a decision-maker (often an authority who orders a bike-sharing service). The method is universal and could be used in differ cities. In the chapter the application of the method is presented based on the case study of the city Krakow in Poland. The city has around 780 981 residents and the 326 sq.km area. Until the end of 2019 in the city operated Wavelo bike-sharing system of station-based type with BikeU company as the main provider.

Besides an introduction part, this chapter consists of 5 sections. The Sect. 2 is a review of the existing method to evaluate urban bike-sharing systems and their functioning. Details concerning performance parameters, which proposed to evaluate urban bike-sharing system, and an approach to assess the attractiveness of bike stations, which is considered as an additional necessary information, have been provided in the Sects. 3 and 4, respectively. Results of case study analysis have been provided in the Sect. 5. The conclusions and discussion was presented in the Sect. 6 together with providing future research plans for the body of problems addressed in the chapter.

2 Literature Review

The literature presents publications describing the operation of the urban bicycle system [13]. The approach to analyse the functioning is differentiated depending on the goal set for the decision-makers, e.g. introduction of a new system, development of the current one, technical changes (transition from a docked system to a partially dockless one). The authors [14] compare the docked and dockless bike-sharing systems, indicate how the systems compete and how they complement each other based on a San Francisco case study. They point out that two systems operating simultaneously in one city have a beneficial effect on the user, who choose bike for the short journeys from the docked system (for specific purposes: work, between intersections) and the dockless system for longer journeys. The dockless system also serves the districts with low population density. It must be mentioned that this article presents only an analysis of the application of two systems in one, which brings the benefits and shows behaviour of users.

The data from the urban bicycle system also allow for the development of accompanying infrastructure such as bicycle paths. The systems are often equipped with a GPS transmitter, which allows to collect accurate data. An example is the work of authors [15], in which GPS data was used to assess the speed of cyclists on different types of bicycle infrastructure. These data are also used to estimate the intensity of bicycle traffic based on information from the city bicycle system [16]. Additionally, it is proposed to verify such data using continuous automatic measurement in each research area, which allows to estimate the share of bikes from city bike-sharing system. The indicated data are also useful in developing a traffic model for the city [17]. The data on the course of GPS routes allow for an accurate representation of the actual situation on the roads in each area. Thanks to the detailed representation of the road network, important indicators such as speed in traffic, average waiting time at intersections and even waiting times for specific turn-by-turn manoeuvres at intersections can be read from the speed profiles. These are important figures for network analysis and calibration of route selection models [18].

The authors [19] draw attention to the need to apply a time-space analysis for the urban bicycle system. It is important to find places with high demand for the proposed services. In the presented case the places with the highest demand are points of interest, subway stations, and dormitories. The authors focus on solutions related to the combination of scooters and bikes. They do not present solutions to assess the functioning of the current system. The idea for strategic planning of the public bicycle system with its service was presented in [20]. The proposed mathematical model considers the interests of users and investors by determining the location of bicycle stations. The specific factors influencing the location of the stations are the distance: between the stations, the origins and destinations of travel, and the costs. The presented method is related to planning the new system. Another method to design an urban bike-sharing system is presented in [21] and is related to the simulation and optimization approach. The application allows optimizing the location of bicycle stations to increase the modal share of bikes. The authors draw attention to the necessity of good access to the stations to increase the number of trips. The analysis did not consider the land use around bicycle stations. The paper [22] presents the way to optimize of bicycle stations location with GIS approach. The potential demand has been considered by means of a spatial distribution of possible demand for journeys with using factors, e.g., density of people, number of workplaces. The method is described from the viewpoint of creating a new system, which does not consider data from the current system. A similar approach from the perspective of the designed systems is presented in [23], where the method based on the maximum area coverage of bicycle station locations is presented considering various factors. The studies presented in [24] indicate the necessity of such a planned bicycle system to be closely related to public transportation.

From the viewpoint of the user of an urban bicycle system, the level of service is considered, which is described by various variables such as: terrain topology, climate, socio-demographic, traffic obstructions, possibility to change the means of transport. In addition, factors related to infrastructure are considered: routes of rented bicycles, law enforcement, movement surfaces, trip-end facilities [13, 25].

3 Methodology

3.1 Bike-Sharing Systems, Their Types and Metrics

The general goals when introducing a bike-sharing system could be following, according to OBIS project partners [26]:

- to promote cycling as a mode of transport;
- to support a public transport system;
- to create new employment opportunities by setting up a bike-sharing system;
- to increase a comfort of tourists;
- to improve an image of city/region;
- to influence positively on the inhabitants' health.

The installation of such system could cover one or more of these aims, which strongly depends on its type and characteristics. Thus, besides the definition of service (coverage) area, the main choice among three possible types of bike-sharing systems should be done: station-based system (docked system), dockless system or hybrid one [27]. In comparison to other types, for station-based systems, which performance is considered and discussed within the scope of this chapter, the location of stations is a critically important aspect. Depending on it, three general concepts could be identified:

- bike-and-ride scheme, which aim is to support a public transport system by building stations near bus/tram stops as well as near interchange points of different levels;
- tourist-oriented scheme, which aim is to increase a comfort of tourists by locating stations close to different points of interests;
- resident-oriented scheme, which aim is to promote cycling among local inhabitants by installing stations near their every-day activities.

To describe bike-sharing systems, guidelines prepared by the Institute for Transportation and Development Policy (ITDP) and published in 2014 and 2018 [27, 28] present following basic system metrics:

- system coverage area;
- population in system coverage area;
- number of bikes;
- number of stations;
- number of docks (parking places for bicycles);
- types of users: casual/long-term.

Guidelines mention also such performance metrics: average daily trips per bike and average daily trips per resident. Such general parameters are good to assess the system when comparing it with another one [29], but not to evaluate the system itself.

With regard to research of bike-sharing system performance, two papers must be mentioned. Zhang and Huang [30] made an analysis of such system in Wuchang Area of Wuhan (China) using users' survey and available data about infrastructure (road network, station locations, etc.). As the survey returns information concerning bike users' profiles and their view on the system, based on the infrastructural data a spatial analysis on bike stations was done to assess their catchment population as well as their accessibility. Borgnat with co-authors [31] published in 2011 very detailed analysis of bike-sharing system in Lyon (France) based on the records of all bicycle trips during more than two years. Taking into account such dataset, analysis has quite a large range: temporal and spatial patterns, individual characteristics of trips, building short- and long-term forecasting models. Also it must be mentioned that this research paper was oriented mainly to describe and to present general trends of the system; as authors wrote in conclusions, operational investigation objectives could not be discussed because there was a lack of qualification for public disclosure.

3.2 Parameters to Evaluate the Performance of Bike-Sharing Systems

Today bicycles of all modern sharing systems independently of their types are equipped with GPS transmitters. They allow service providers not only to establish current positions of bikes in the coverage area, which creates an opportunity to react fast on critical situations as well as to realize an effective management of the whole system, but also to gather information about movement of these bicycles and their users. GPS data could be collected in the form of GPS traces of bike trips: during each trip, geographic coordinates together with information about day and time are recorded each few seconds (for example, each 5 seconds). In spite of the fact that the availability of such GPS data are restricted, there are already an amount of research works based on such type of dataset [31, 32]. Further basic assumptions concerning data from station-based bike-sharing systems are described below when discussing different performance parameters.

As it was mentioned above, ones of basic metrics of each system are the number of stations $\boldsymbol{M}$ and the number of bikes $\boldsymbol{K}$. Each station could be described by using following parameters: besides its number, station name, its geographic coordinates (latitude and longitude), number of parking places for bicycles (docks).

Each trip ***n*** made by bike ***k*** from station ***i*** to station ***j*** could be characterized with the help of following information:

- day when this trip was done (a rare exception could arise when it begins in one day and ends on the next one);
- moment of time when this trip begins – t_{nki};
- moment of time when this trip ends – t_{nkj};
- trip length l_{nkij};
- user characteristics (e.g., age, gender).

It must be mentioned that length of each bike trip could be established based on a GPS trace of this trip, which gives an opportunity to get quite exact distance value. A poor alternative could be the calculation based on coordinates of trip begin and end together with the application of appropriate metrics.

N_{kij} is an amount of all trips made by bicycle ***k*** from the station ***i*** (start of trip) to station ***j*** (end of trip) at a certain time (hour, day, week, month, etc.).

Based on all above-mentioned features, the amount of performance parameters could be established. Firstly, the load of station ***i*** as an origin of trips is calculated according to the Eq. 1:

$$LO_i = \sum_{k=1}^{K} \sum_{j=1}^{M} N_{kij} \tag{1}$$

The second parameter is the load of station ***j*** as a destination of trips established according to the Eq. 2:

$$LD_j = \sum_{k=1}^{K} \sum_{i=1}^{M} N_{kij} \tag{2}$$

The general load of station ***i*** as both origin and destination of trips is calculated per the Eq. 3:

$$GL_i = LO_i + LD_i \tag{3}$$

Travel time of trip ***n*** from the station ***i*** to the station ***j*** by bike ***k*** is determined with the help of the Eq. 4. It is important here to differ travel time when user rides a bike and usage time between moments when user unlocks a bike when starting its usage and locks it when ending.

$$T_{nkij} = t_{nkj} - t_{nki} \tag{4}$$

The average length of trips from the selected station ***i*** made by all bicycles is established by Eq. 5:

$$l_{\text{average i}} = \sum_{k=1}^{K} \sum_{j=1}^{M} \frac{\sum_{n=1}^{N_{kij}} l_{nkij}}{N_{kij}} \tag{5}$$

The average length of trips to the selected station ***j*** made by all bicycles is calculated according to Eq. 6:

$$l_{\text{average j}} = \sum_{k=1}^{K} \sum_{i=1}^{M} \frac{\sum_{n=1}^{N_{kij}} l_{nkij}}{N_{kij}} \tag{6}$$

The number of trips made by bike ***k*** in a certain time period is determined by using Eq. 7:

$$R_{\text{k}} = \sum_{i=1}^{M} \sum_{j=1}^{M} N_{kij} \tag{7}$$

The total travel time ridden by bike **k** in a certain time period is calculated with the help of Eq. 8:

$$T_{\text{k}} = \sum_{i=1}^{M} \sum_{j=1}^{M} \sum_{n=1}^{N_{kij}} T_{nkij} \tag{8}$$

Average travel time by bike **k** over a certain period of time is established by Eq. 9:

$$T_{\text{average k}} = \frac{T_k}{R_k} \tag{9}$$

All above-presented parameters can be used to evaluate the basic performance of bike-sharing systems for the viewpoint of its basic elements: stations, bikes as well as its main outputs – trips. Only parameters of users are not considered here, because in the most cases the sensitivity of personal data restricts its availability: there is not a lot of examples when such data were available in a wide scale to make them public (e.g., bike-sharing system provider in New York opened up genders and ages of registered users for each trip).

4 Attractiveness of Bike-Sharing Stations

As it was mentioned above, the location of stations is a critically important for bike-sharing systems of station-based type. According to ITDP guidelines from 2014 and 2018 [27, 28], there is an amount of facts influencing on it. First of all, it must be pointed out the preferability to locate stations close to public transport stops, near points generated different type of activities, close to bicycle infrastructure, etc.

Taking into account above-presented information, it was decided to develop simple scales to assess the attractiveness of each station based on its location surrounding. As activities were taken standard groups of everyday trip motivations such as "house" to include a living area potential, "work" – a potential of working places, "education" – a potential of education facilities, "shopping" – a potential of shops and malls. Also an activity "leisure time" was considered, which included tourist attractions and points of interest. All these features were assessed with the help of specially developed 10-point scales for each activity depending on catchment areas of stations for distances up to 300 m and 300–500 m (for the activity "leisure time" – also distances 500–800 m was taken into account). Details of all activities and their attractiveness points are shown in Tables 1, 2, 3, 4 and 5.

Additionally, criterion "public transport" was considered, which is based on the accessibility of public transport stops and interchange points (Table 6). The assessment of its attractiveness was done similar to ideas as used to estimate points for

Table 1 Activity "house" and its attractiveness points

Type of building system	Distance to the bike-sharing station	
	Up to 300 m	300–500 m
Low-rise, low-density buildings	1	0
Low-rise, high-density buildings	2	1
Mid-rise, low-density buildings	4	2
Mid-rise, high-density buildings	6	3
High-rise, low-density buildings	8	5
High-rise, high-density buildings	10	7

Table 2 Activity "work" and its attractiveness points

Density of working places	Distance to the bike-sharing station	
	Up to 300 m	300–500 m
low-density of working places	1	0
Mid-density of working places	2	1
High-density of working places	4	2
Office building	8	5
Office complex	10	7

Table 3 Activity "education" and its attractiveness points

Education facilities	Distance to the bike-sharing station	
	Up to 300 m	300–500 m
1–2 primary schools/junior high schools	1	0
3 and more primary schools/junior high schools	2	1
1–2 senior high schools	4	2
3 and more senior high schools	6	4
University	10	7

Table 4 Activity "shopping" and its attractiveness points

Shopping facilities	Distance to the bike-sharing station	
	Up to 300 m	300–500 m
Small shops	2	0
Supermarket	4	1
Small shopping center	6	2
Middle or large shopping mall	10	5

Table 5 Activity "leisure time" and its attractiveness points

Type of recreation/free time attractions	Distance to the bike-sharing station		
	Up to 300 m	300–500 m	500–800 m
Tourist points of interest	10	8	6
Low density of restaurant/pub	4	2	0
High density of restaurant/pub	8	5	3
Cinema/theater	4	2	0
Popular leisure areas	10	9	6
Areas near rivers/lakes	6	4	2

Table 6 Criterion "public transport" and its attractiveness points

Public transport accessibility	Distance to the bike-sharing station	
	Up to 300 m	300–500 m
Bus stop	2	1
Tram stop	4	2
Bus-bus interchange point	6	4
Tram-bus interchange point	8	6
Tram-tram interchange point	10	8

above-mentioned activities. Infrastructure criteria were not used within the scope of this analysis.

5 Case Study

5.1 Bike-Sharing Systems in Krakow

Bike-sharing system in Krakow has a long history. The first system was introduced in 2008 and started to operate at full power from 01.04.2009. It was also the first bike-sharing system in Poland. During 10 years of its existence the service area as well as amounts of stations and bicycles were changing a lot. According to official information, the maximal number of operating stations was 169. This achievement was reached in March 2019. The maximal amount of bikes in use were 1500 (October 2017). Unfortunately, the system was taken out of service in the end of 2019, because the city authority did not find a new service provider. The last name of the system was Wavelo, which was used from 2016 till 2019.

This research work is based on GPS traces from the bike-sharing system in Krakow, which covers the period from 1st June till 30th June of the year 2017 and was furnished by service provider. Traces were generated by GPS transmitters of Wavelo bikes, which were to read the bike's position at intervals of 5 seconds, to write this information with the date and time of the reading and to save this data in GPX files. To extract a necessary information, such files was parsed using a special program written by the authors by using Python programming language. Having such data, it was assumed that one GPS trace was considered as an information about one bicycle trip.

Taking into account that Wavelo system allowed to leave a bicycle out of stations by paying an additional fee, the first step was to eliminate all such trips. This work was done simultaneously together with determining the location of bicycle stations in the considered period, which was based on photos from the Google Street View service from the same period of time. In such way it was estimated that 149 stations operated in Krakow in June 2017. Their locations are shown in Fig. 1. As a result, analysed database consisted of 133,527 records, which included only trips between stations – around 89% of all available GPS traces.

The main restrictions of the available dataset were a certain amount of shortcomings. The main of them was a lack of bicycle identification: trips could not be assigned to a certain bike. In such situation, it was impossible to analyse their performance. The second one, which could be mentioned, is a lack of basic users' data (age and sex), but as it was pointed above, its availability is a common problem of many bike-sharing systems.

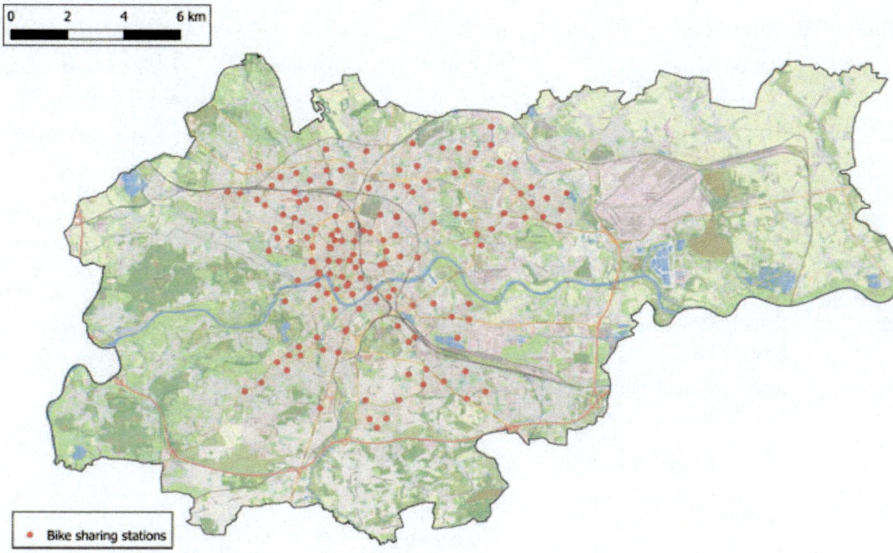

Fig. 1 Location of bike-sharing stations in Krakow (the state on June 2017)

5.2 Balance of the Bike-Sharing System in Krakow

Frist of all, it must be pointed out that the system was generally in balance when comparing the difference of origin and destination loads for each stations: only 30.9% of stations have this parameter more than 10% and only 10.7% – more than 20%. It means that there was no necessity to move bicycles between locations manually.

As special cases, the so-called double stations must be considered. Under them are understood stations, which are situated from each other for less than 250 m. Wavelo bike-sharing system had 7 such double stations. Their names together with their parameters are presented in the Table 7. The most critical situation looks for group 3, where disbalance reaches 85%. It could be explained by the fact that the station "Cracovia Błonia" with the higher load was situated to a great extent on the crossroad of popular walking routes in comparison to its close neighbour, which attracted more attentions of users. The second critical case is stations near the Bagatela Theatre. There is a large difference in the general load towards the station, which was situated closer to the city center ("Teatr Bagatela/Szewska"), though the station "Teatr Bagatela/Karmelicka" was wining in regards of public transport accessibility. This fact seems to be surprising, because, as it will be discussed below, the bike-sharing system in Krakow is more public transport orientated. Also it is interesting that the both extreme cases refer to double stations with ones of the smallest distances on foot between each other.

Table 7 Double stations

Group	Station name	Coordinates	Distance between stations [m]	Difference in general loads of stations	
					[%]
1	Jubilat	50.055501, 19.927601	80	144	6.8
	Kościuszki	50.05499, 19.927522			
2	Rondo Matecznego/ Rydlówka	50.036571, 19.939955	110	293	17.3
	Rondo Matecznego/ Kalwaryjska	50.037165, 19.940571			
3	Cracovia Błonia	50.059188, 19.923683	60	3193	85.1
	Muzeum Narodowe/Focha	50.059151, 19.924396			
4	Teatr Bagatela/Szewska	50.063254, 19.933348	65	1450	25.2
	Teatr Bagatela/Karmelicka	50.063564, 19.932736			
5	Stradom/Rossmann	50.051994, 19.941847	130	268	21.1
	Stradom	50.051183, 19.941013			
6	Rondo Grzegórzeckie	50.058114, 19.959718	180	383	14.7
	Grzegórzecka	50.05775, 19.95771			
7	Narzymskiego 1	50.073532, 19.965862	50	243	25.0
	Narzymskiego 2	50.073088, 19.96592			

5.3 *Distance to the City Center as a Critical Parameter*

Based on general loads of the stations and with the help of a simple spatial analysis, it was established the importance of distances from bike-sharing stations to the city center of Krakow: a station load increases if this station is closer to the center. To determine this fact, the city area was divided into squares with size 500 × 500 m. Then loads of all stations were assigned to these squares depending on their localization. The result of this action is shown on the Fig. 2.

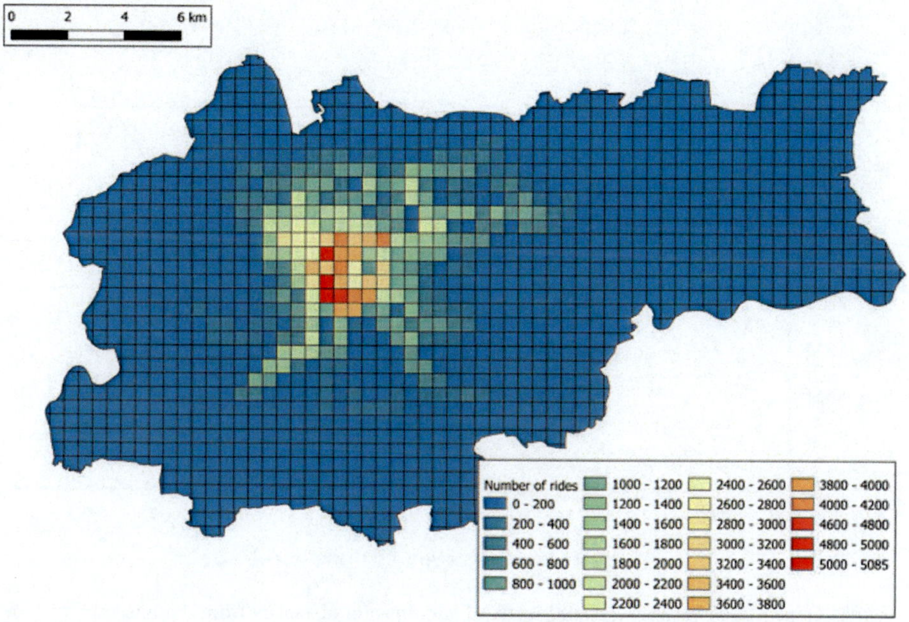

Fig. 2 Spatial location of general loads for bike-sharing stations in Krakow

5.4 Performance Parameters as a Key to Understand a Bike-Sharing System

To evaluate Wavelo bike-sharing system, following performance parameters were calculated: general load of each station, average length of trips from and to each station, attractiveness points of each station, distance by bike from each station to city center. According to the preliminary analysis based on these parameters, it was observed a clear bias of Wavelo system towards public transport orientation, which allows to classify it to the bike-and-ride scheme. Further analysis will be done separately for stations with different attractiveness points of the criterion “public transport”.

The first group includes 18 stations got 10 points for public transport accessibility. In Fig. 3 is presented the dependence between general loads of stations and distances from them to city center by bike for such stations. Parabolic trend line shows quite a high correlation ($R^2 = 0.75$). It explains very clear a tendency: besides a logical decrease of a general load when increasing the distance to the center, there is a moment when stations are so far that they start to search another “center” or to generate the own one. According to the diagram, the critical distance for Wavelo system was around 7.6 km.

Table 8 presents stations, which crossed the critical distance to the city center. Their distances to the city center are quite similar as well as the differences in the total attractiveness points are not so high, but average trip lengths and general load

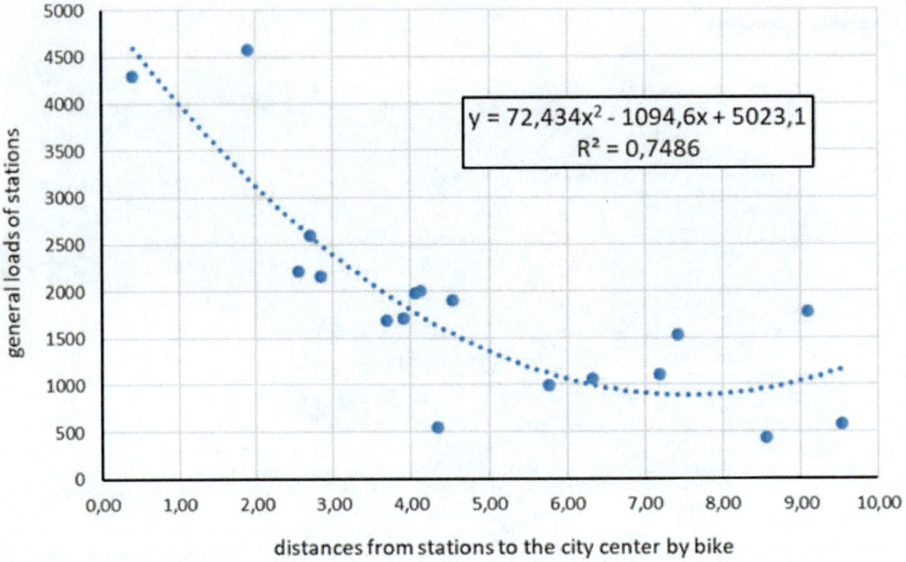

Fig. 3 Dependence between general loads of stations and distances from them to the city center by bike for stations with 10 points for the criterion "public transport"

Table 8 Stations with 10 points for the criterion "public transport", which crossed the critical distance

Station name	Coordinates	Total points	Average trip length	Distance to center by bike	General load of station
Mistrzejowice	50.095102, 19.99674	28	4.4	8.56	439
Plac Centralny	50.072739, 20.037646	32	3.9	9.09	1773
Osiedle Bohaterów Września	50.101751, 20.012794	29	5.4	9.54	584

shows that the station "Plac Centralny" seems to generate a separate center, which confirms also by the status of this location as a center of Nowa Huta district as well as by high attractiveness points for "leisure time" activity.

The second group consists of 18 stations, which have 6–8 points for public transport accessibility. Figure 4 presented the similar dependence for this group as for stations of the first group, but with smaller correlation (R2 = 0.56), which means an increasing influence of another factors on their loads. Also it must be mentioned that two stations from double stations groups – "Stradom/Rossmann" and "Muzeum Narodowe/Focha" – were eliminated from the diagram because of their clear disbalance in the general loads discussed above.

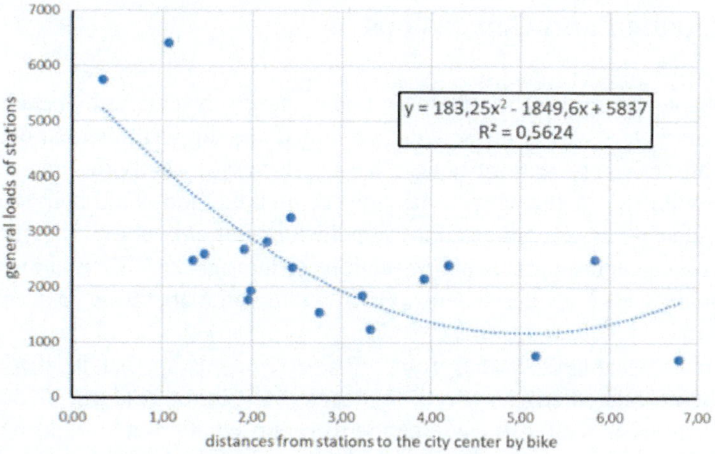

Fig. 4 Dependence between general loads of stations and distances from them to the city center by bike for stations with 6–8 points of the criterion "public transport"

The third group includes stations with 2–4 points for public transport accessibility. Table 9 presents the summary of performance parameters for these stations. Subgroups were divided based on average trip length: more than 4 km – subgroups 1 and 2, less than 4 km – subgroups 3, 4, 5 and 6 – and on average load of stations: 0 to 1000, 1001–2000, 2001–3000 and more than 3001. Besides the clear correlation between general loads and distance to the city center, there is an increasing dependency between loads and total attractiveness points.

Table 9 Stations of the third group with 2–4 points for the criterion "public transport"

Subgroups	Number of stations	Public transport points[a]	Total points[a]	Trip length[a]	Distance to center by bike[a]	General load of station[a]	General load of station (range)
1	25	3.1 (1.0)	10.2 (2.7)	4.7 (0.7)	7.3 (1.8)	666 (235)	78–979
2	23	3.0 (1.0)	18.9 (5.6)	4.4 (0.4)	5.9 (1.6)	1402 (288)	1006–2006
3	11	3.5 (0.9)	19.1 (6.9)	3.7 (0.5)	6.2 (3.3)	796 (133)	517–973
4	24	2.8 (1.0)	19.2 (4.9)	3.3 (0.4)	2.6 (1.2)	1507 (304)	1001–1980
5	20	3.5 (0.9)	23.0 (5.7)	3.4 (0.4)	2.5 (1.3)	2467 (283)	2124–2905
6	8	3.5 (0.9)	24.8 (5.2)	3.3 (0.4)	1.6 (0.9)	4083 (802)	3056–5500

[a]Average values for sub-groups with the standard deviation in parentheses

6 Discussion and Conclusions

In this chapter an approach to evaluate a bike-sharing system with the help of performance parameters was described and explained. To support this method, criteria of stations attractiveness was introduced based on everyday activities and public transport accessibility. As the case study showed, such method could help to understand the bike-sharing system and to draw conclusions not only about the whole system, but also about each separate sharing stations or their groups. Such knowledge could help to make well-considered decisions concerning the operation or development of the system.

The main weakness of the presented case study is so that it is oriented on a station-based bike-sharing system. It means that further analysis of dockless and hybrid systems as well as the adaptation of proposed approach or, probably, search of an appropriate method are required. The next aspect for further development of this topic is to turn to another type of system concept, for example, oriented on tourists, which are quite actual, especially, for coastal cities [33].

Attractiveness of bike-sharing stations could be considered as another weak point of this approach. On the one hand, the simplest scales were chosen to define points for each stations, which could be applied manually, on the other hand, there is a large amount of special GIS software packages, which could support such process making it more precise. Also in such situation there is a possibility to use such traditional approach as questionnaire to understand behaviour and choice of bike users [36, 37]. Additionally, transport infrastructure could be easily taken in the consideration [34]. Such data could be improved also by deep analysis of social sensing of bike-sharing system users [35].

Looking at the experience in Poland, it could be mentioned that local authorities of cities or regions include in a contract mainly the amount of stations and bikes as basic system parameters when ordering the introduction and maintenance of bike-sharing system in their area. That is why detailed analysis of the bike-sharing system performance or at least the calculation of basic performance parameters is most commonly not a part of contractors' obligations and, therefore, has not been carried out in most cases, although such actions could bring benefits as described in this chapter. Thus, each service provider organizes processes of data recording and storing according to its needs and own data collection policy, which results often in a lack of necessary data to make the full analysis of system performance or any further analysis or research. In some cases, problem of data availability covers with data sensitivity, especially, when it is referred to users.

References

1. Klöckner CA, Nayum A, Mehmetoglu M (2013) Positive and negative spillover effects from electric car purchase to car use. Transp Res Part D: Transp Environ 21:32–38
2. Sendek-Matysiak E (2019) Multi-criteria analysis and expert assessment of vehicles with different drive types regarding their functionality and environmental impact. Sci J Silesian Univ Technol Series Transp, vol 102, pp 185–195
3. Sándor Zs, Csiszár Cs (2015) Modelling and analysis methods of integrated information systems of transportation. In: 4th International conference on models and technologies for intelligent transportation systems (MT-ITS 2015), 3–5 June 2015, pp 348–355. Budapest, Hungary. https://doi.org/10.1109/MTITS.2015.7223278
4. Sierpiński G, Macioszek E (2020) Equalising the levels of electromobility implementation in cities. In: Mikulski J (ed) Research and the future of telematics, communications in computer and information science, vol 1289. Springer, Heidelberg, pp 165–176
5. Macioszek E, Sierpiński G (2020) Charging stations for electric vehicles—Current situation in Poland. In: Mikulski J (ed) Research and the future of telematics, communications in computer and information science, vol 1289. Springer, Heidelberg, pp 124–137
6. van Waes A, Farla J, Raven R (2020) Why do companies' institutional strategies differ across cities? A cross-case analysis of bike sharing in Shanghai & Amsterdam. Environ Innov Societal Transit 36:151–163
7. Kłos MJ (2016) Estimation of effects caused by the implementation of Park&Ride system in the transport hub. Transp Probl 11:5–12
8. Chen F, Turoń K, Kłos M, Czech P, Pamuła W, Sierpiński G (2018) Fifth-generation bike-sharing systems: examples from Poland and China. Sci J Silesian Univ Technol Series Transp 99:05–13. https://doi.org/10.20858/sjsutst.2018.99.1
9. Sierpiński G, Staniek M, Kłos MJ (2020) Decision making support for local authorities choosing the method for siting of in-city EV charging stations. Energies 13(18):4682. https://doi.org/10.3390/en13184682
10. Kabak M, Erbaş M, Çetinkaya C, Özceylan E (2018) A GIS-based MCDM approach for the evaluation of bike-share stations. J Clean Prod 201:49–60. https://doi.org/10.1016/j.jclepro.2018.08.033
11. Cheng P, OuYang Z, Liu Y (2019) Understanding bike sharing use over time by employing extended technology continuance theory. Transp Res Part A: Policy Pract 124:433–443. https://doi.org/10.1016/j.tra.2019.04.013
12. Mooney SJ, Hosford K, Howe B, Yan A, Winters M, Bassok A, Hirsch JA (2019) Freedom from the station: spatial equity in access to dockless bike share. J Transp Geogr 74:91–96. https://doi.org/10.1016/j.jtrangeo.2018.11.009
13. Kazemzadeh K, Laureshyn A, Winslott Hiselius L, Ronchi E (2020) Expanding the scope of the bicycle level-of-service concept: a review of the literature. Sustainability 12(7):2944. https://doi.org/10.3390/su12072944
14. Lazarus J, Pourquier JC, Feng F, Hammel H, Shaheen S (2020) Micromobility evolution and expansion: understanding how docked and dockless bikesharing models complement and compete–A case study of San Francisco. J Transp Geogr 84: https://doi.org/10.1016/j.jtrangeo.2019.102620
15. Kieć M, Pogodzińska S (2018) Assessment of speed for various types of bicycle infrastructure based on GPS data [in Polish: Ocena prędkości rowerzystów na różnych typach infrastruktury rowerowej z wykorzystaniem danych GPS]. Transp Miejski i Regionalny 4:20–24
16. Pogodzińska S (2018) Assessment of speed for various types of bicycle infrastructure based on GPS data [in Polish: Szacowanie natężenia ruchu rowerowego na podstawie danych z systemu rowerów miejskich]. Transp Miejski i Regionalny 1:5–11
17. Kłos MJ, Sobota A, Żochowska R, Soczówka P (2019) Traffic measurements for development a transport model. In: Sierpiński G (eds) Smart and green solutions for transport systems TSTP 2019. Advances in intelligent systems and computing, vol 1091. Springer. Cham, pp 265–278. https://doi.org/10.1007/978-3-030-35543-2_21

18. Rupi F, Poliziani C, Schweizer J (2020) Analysing the dynamic performances of a bicycle network with a temporal analysis of GPS traces. Case Stud Transp Policy 8(3):770–777. https://doi.org/10.1016/j.cstp.2020.05.007
19. Zhu R, Zhang X, Kondor D, Santi P, Ratti C (2020) Understanding spatio-temporal heterogeneity of bike-sharing and scooter-sharing mobility. Comput Environ Urban Syst 81: https://doi.org/10.1016/j.compenvurbsys.2020.101483
20. Lin JR, Yang TH (2011) Strategic design of public bicycle sharing systems with service level constraints. Transp Res Part E: Logist Transp Rev 47(2):284–294. https://doi.org/10.1016/j.tre.2010.09.004
21. Romero JP, Ibeas A, Moura JL, Benavente J, Alonso B (2012) A simulation-optimization approach to design efficient systems of bike-sharing. Procedia—Soc Behav Sci 54:646–655. https://doi.org/10.1016/j.sbspro.2012.09.782
22. García-Palomares JC, Gutiérrez J, Latorre M (2012) Optimizing the location of stations in bike-sharing programs: a GIS approach. Appl Geogr 35(1–2):235–246. https://doi.org/10.1016/j.apgeog.2012.07.002
23. Frade I, Ribeiro A (2015) Bike-sharing stations: A maximal covering location approach. Transp Res Part A: Policy Pract 82:216–227. https://doi.org/10.1016/j.tra.2015.09.014
24. Böcker L, Anderson E, Uteng TP, Throndsen T (2020) Bike sharing use in conjunction to public transport: exploring spatiotemporal, age and gender dimensions in Oslo, Norway. Transp Res Part A: Policy Pract 138:389–401. https://doi.org/10.1016/j.tra.2020.06.009
25. Fitch DT, Handy SL (2018) The Relationship between experienced and imagined bicycling comfort and safety. Transp Res Record: J Transp Res Board 2672:116–124. https://doi.org/10.1177/0361198118787635
26. Optimising Bike Sharing in European Cities—A Handbook. OBIS, June 2011, p 90. https://ec.europa.eu/energy/intelligent/projects/sites/iee-projects/files/projects/documents/obis_handbook_en.pdf
27. The Bikeshare Planning Guide. Institute for Transportation and Development Policy (2018) https://www.transformative-mobility.org/assets/publications/The-Bikeshare-Planning-Guide-ITDP-Datei.pdf
28. The Bikeshare Planning Guide. Institute for Transportation and Development Policy (2014) https://itdpdotorg.wpengine.com/wp-content/uploads/2014/07/ITDP-Bike-Share-Planning-Guide-1.pdf
29. Bieliński T, Kwapisz A, Ważna A (2019) Bike-sharing systems in Poland. Sustainability 11(9):2458. https://doi.org/10.3390/su11092458
30. Zhang Y, Huang Z (2012) Performance evaluation of bike sharing system in Wuchang area of Wuhan. China. In: 2012 6th International Association for China planning conference (IACP), 13135901. IEEE Press. Wuhan, China. https://doi.org/10.1109/IACP.2012.6342972
31. Borgnat P, Abry P, Flandrin P, Robardet C, Rouquier J-B, Fleury E (2011) Shared bicycles in a city: a signal processing and data analysis perspective. Adv Complex Syst 14(3):415–438. https://doi.org/10.1142/S0219525911002950
32. Albiński S, Fontaine P, Minner S (2018) Performance analysis of a hybrid bike sharing system: A service-level-based approach under censored demand observations. Transp Res Part E: Logist Transp Rev 116:59–69. https://doi.org/10.1016/j.tre.2018.05.011
33. Bakogiannis E, Vassi A, Christodoulopoulou G, Siti M (2018) Bike sharing systems as a tool to increase sustainable coastal and maritime tourism. the case of Piraeus. Region Sci Inquiry 10(3):57–70
34. Pashkevich A (2014) Realization of OpenStreetMap project possibilities: special GIS software described by using case study of transport infrastructure. In: 14th International multidisciplinary scientific GeoConference SGEM 2014, www.sgem.org, SGEM2014 conference proceedings. SGEM, Albena, Bulgaria, pp 799–806. https://doi.org/10.5593/SGEM2014/B23/S11.102
35. Pashkevich A, Karamshuk D (2014) Understanding and improving urban cycle networks with social sensing. In: 34th Conference on transportation systems with international participation Automation in transportation. KoREMA, Zagreb, Croatia, pp 45–49

36. Schweizer J, Meggs JN, Dehkordi NR, Rupi F, Pashkevich A (2012) Unified, low-cost analysis framework for the cycling situation in cities. Int J Civil Environ Eng 6(5):356–363
37. Schweizer J, Meggs JN, Dehkordi NR, Rupi F, Pashkevich A (2012) An analysis framework to assess automobile, public transport and bike usage in west and eastern Europe. In: Proceedings of the 15th International Conference on Transport Science ICTS 2012 – Maritime, transport and logistics science, 28 May 2012, Portorož, Slovenia

A Methodological Approach to the Real-Time Data Analysis from the ViaTOLL System

Vitalii Naumov, Andrzej Szarata, and Hanna Vasiutina

1 Introduction

Traffic data can be used in various spheres, such as economical and statistical studies, low-making, urban planning, and other areas. This data is provided by different traffic measuring systems (collected with the use of cameras, sensors, radars, etc.) and contains information on vehicles' flow characteristics (speed, the flow intensity, and structure), travel times, congestions, implemented traffic management measures. In transportation science, traffic data is used for transport systems modelling, traffic monitoring, development of intelligent transportation systems, etc. The accurate data collection, its analysis, and further use in traffic management systems can significantly improve traffic conditions, increase road safety, and, overall, positively impact the sustainable development of the transport system (e.g. [1–4]).

This chapter aims to present the method for processing of data from the traffic toll system. Our approach allows real-time analysis of information, its results may be used for statistical purposes, as well as initial data for traffic management systems.

V. Naumov (✉) · A. Szarata · H. Vasiutina
Cracow University of Technology, Str. Warszawska 24, 31155 Kraków, Poland
e-mail: vnaumov@pk.edu.pl

A. Szarata
e-mail: aszarata@pk.edu.pl

H. Vasiutina
e-mail: hanna.vasiutina@doktorant.pk.edu.pl

G. Sierpiński and E. Macioszek (eds.), *Decision Support Methods in Modern Transportation Systems and Networks*, Lecture Notes in Networks and Systems 208,
https://doi.org/10.1007/978-3-030-71771-1_8

2 Literature Review

Procedures for gathering, analysing, and predicting road traffic parameters concern a vast number of scientists, traffic operators, and engineers. The source of these parameters is data collected from intelligent transportation systems These systems use different technologies for vehicle detection: toll fee collection system [5–15], road sensors [16], Bluetooth [17], mobile data [18], inductive loops [19], and others.

The taxonomy of the traffic forecasting literature was conducted by [5]. Among other things, the authors distinguished three main groups of detectors: point (which out-put traffic flow, occupancy, and density), interval (produce traffic measures between road points), and moving sensors (which provide individual vehicle data). Toll collection systems belong to the interval data source and can be of three types: manual (with the operating collector in a booth), semi-automatic (vehicle stops until the gate opens after a toll card is scanned), and fully automatic (vehicle don't have to stop due to wireless transaction at the gate).

Data from toll systems are widely used for the optimization of transport systems. Projects, that use such data, mainly focus on two main goals: traffic management enhancement and improvement of the toll system itself. Besides, since the real-world data contains noise, errors and may lead to inaccurate outcomes, authors propose a set of methods and techniques for obtaining, processing, and modifying such data [6]. The methods for data manipulation were also described by [16] in order to achieve accurate results on the traffic intensity. For this study, such traffic characteristics as average speed and vehicle counts were gathered from road sensors.

The electronic toll collection data served as an input in the dynamic origin-destination model to estimate the traffic volumes [7]. Moreover, the analysis of achieved results helped authors to explore traffic demand and its variation in order to improve network management. Another approach for estimating travel times on closed toll highways, that aims to improve traffic operations, is presented by [8]. As well as distance-based toll optimization framework was introduced by [9], where authors evaluate the dependence of a toll price and travel parameters by using clustering techniques. Furthermore, the comparison of different techniques for traffic prediction models was conducted by [10]. The authors aim to improve toll-plaza operations and observed the best performance in the K Nearest Neighbours method. The study of [11] explored the influence of the toll removal on the choice of destination using a mixed logit model. Traffic calculation before and after toll removal showed that the trip pattern didn't change significantly. Also, a practical method was developed by [12] to receive the parameters of the traffic flow by retrieving the toll data.

The research [13] on increasing the effectiveness of the toll collection systems presents optimization algorithms and traffic flow prediction models aiming at travel times and queue reduction on the toll plaza. Another study [14] is focused on comparing cash and electronic toll system by evaluating such characteristics as transaction time and capacity of the tollgate. A different approach, that explores various scenarios of servicing at the toll plaza by implementing a microsimulation model in PTV Vissim was described by the paper [15].

Other sources of data, such as Bluetooth [17] and mobile phone [18], were used to achieve vehicular trajectories and vehicle travel times and congestion detection respectively. Characteristics, that describe the individual vehicles in traffic, were obtained by [19] with help of measuring systems equipped with inductive loops and piezoelectric sensors.

3 Description of the Analysed Data Set

The analysed data set contains the parameters of all transactions generated in the roadside infrastructure of the viaTOLL system in the period between January 1, 2019 (the time of the first transaction is 2018/12/31 23:27:05) and January 13, 2019 (the time of the last transaction is 2019/01/13 23:48:38). The number of transactions in the analysed time period is uneven in terms of the day of the week (see Fig. 1): on holidays (2019/01/01), Saturdays (2019/01/05, 2019/01/12), and Sundays (2019/01/06, 2019/01/13), the number of transactions in the viaTOLL system is significantly lower. The daily number of transactions in the analysed set ranges from 376,258 (2019/01/01) to 4,035,078 (2019/01/09). The total number of transactions in the dataset is approximately 35 million of records.

One record contains the following transaction characteristics:

- unique transaction identifier,
- transaction type (0—transaction in roadside infrastructure, 1—next toll collection),
- unique vehicle identifier generated based on the on-board device identifier,
- travel time through roadside infrastructure,

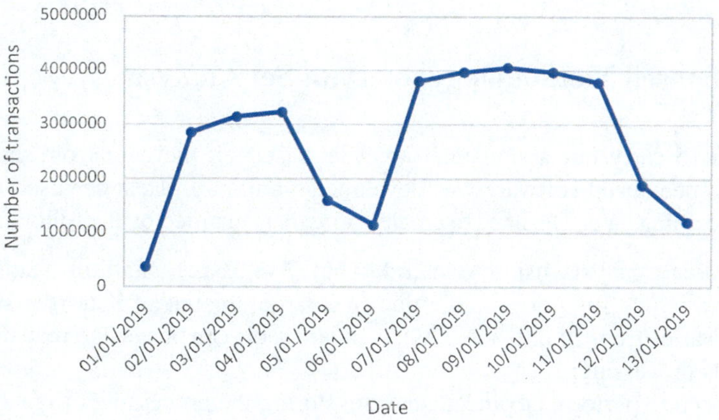

Fig. 1 Dynamics of changes in the number of transactions in the analysed period

Table 1 Vehicle class coding

Class code	Characteristics of the vehicles
13	Light vehicles with GVM and GTM equal to or less than 3.5 tons
14	Light vehicles with GVM equal to or less than 3.5 tons, capable of towing a trailer
30	Coaches with more than 9 seats (including the driver)
41	Heavy goods vehicles with GVM and GTM over 3.5 tons and less than 12 tons
42	Heavy goods vehicles with GVM over 3.5 tons and less than 12 tons with the ability to tow a trailer
50	Heavy goods vehicles with GVM and GTM over 12 tons

- date and time stored as part of the transaction,
- unique identifier of the toll road section,
- unique identifier of the toll station that generated the transaction,
- unique identifier of the last preceding toll station,
- vehicle class code,
- Euro emission class.

The coding of the vehicle class in the analysed data set is performed as shown in Table 1.

Additionally, the analysed data set is supplemented with information about the viaTOLL system gantries. The records with information about gantries correspond to the system gates and have the following structure:

- unique name of the gantry in the system,
- unique identifier of the toll collection station,
- longitude (GPS coordinate in decimal format),
- latitude (GPS coordinate in decimal format).

The total number of gantries in the viaTOLL system for the analysed set is 1,888.

4 Proposed Methodology for Data Set Analysis

In order to carry out a statistical analysis based on the data from the viaTOLL system, specialized software was developed—a library of classes used for reading and processing data. The need to create software is justified by the following reasons:

- statistical analysis using standard tools (MS Excel, StatSoft Statistica, StatGraphics, etc.) is not possible due to a large amount of data (the size of data files varies from 34.3 to 366.7 M): standard tools can be used to read data, but not for its processing;
- the need to perform advanced analyses during the processing of raw data, which cannot be performed using the functions of standard tools.

The viaTOLL data analysis software was developed in Python as the leading technology currently used to develop specialized tools for big data analysis. The following classes have been developed within the specialized library:

- *Gate*: the viaTOLL system gantries within the data model,
- *Transaction*: in the data model transactions occurring in the system,
- *Vehicle*: heavy goods vehicles performing their trips within the viaTOLL system,
- *DataModel*: the data model, containing the basic methods of analysing the data set from the viaTOLL system.

The structure of the developed classes is presented in the form of a UML-diagram in Fig. 2.

The fields of the Gate class reflect the most important characteristics of the viaTOLL system gantries:

- *name*: contains a unique name of the gantry,
- *id*: stores a unique identifier of the toll station,
- *longitude* and *latitude*: contain the gate geographic coordinates.

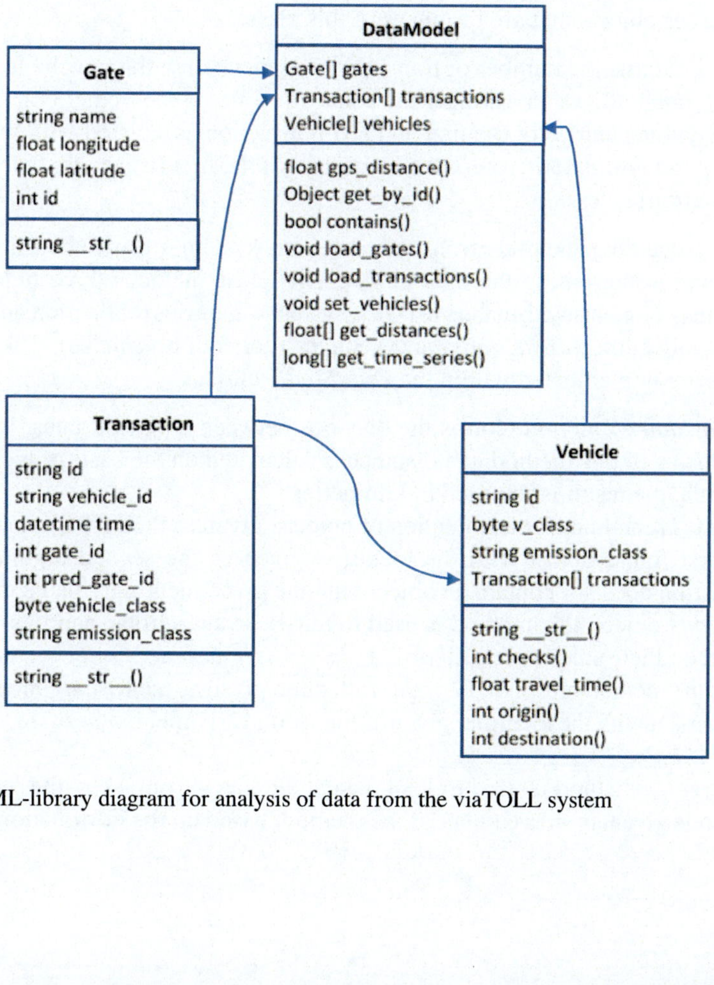

Fig. 2 UML-library diagram for analysis of data from the viaTOLL system

The fields of the *Transaction* class allow to describe the main parameters of a single transaction in the system:

- *id*: contains a unique transaction identifier,
- *vehicle_id*: contains the identifier of the vehicle associated with the transaction,
- *time*: stores the time moment when the vehicle passed through the gantry,
- *gate_id* and *pred_gate_id*: contain the identifiers of the current and previous toll stations,
- *vehicle_class* and *emission_class*: store the vehicle class codes and the Euro emission class respectively for the vehicle associated with the transaction.

The *Vehicle* class has the following fields:

- *id*: a unique identifier of the vehicle in the system,
- *v_class* and *emission_class*: contain the codes of the vehicle class and the Euro emission class,
- *transactions*: a list of all transactions associated with a given vehicle for the analysed data set.

The *Vehicle* class additionally has the following methods implemented as properties of objects that are instances of this class:

- *checks*: returns the number of transactions associated with the vehicle,
- *travel_time*: allows to estimate the total trip time of the vehicle on the section between the gantries of the first and last transaction associated with the vehicle,
- *origin* and *destination*: return the identifiers of the first (trip source) and last (trip destination) gates.

Data processing methods from the viaTOLL system are included in the *DataModel* class representing the data model. This class includes three fields that are collections of objects within the viaTOLL system—gantries (collection *gates*), transactions (collection *transactions*) and vehicles (collection *vehicles*). The following methods are implemented within the *DataModel* class:

- *gps_distance* function returns the distance between a pair of gates, indicated as arguments of this method; the distance is calculated on the basis of the Haversine formula, the result is returned in kilometres.
- *get_by_id* method: for a collection of objects, given as the first argument, returns the first found object with the identifier given as the second argument; if the collection does not contain an object with the given identifier, the method returns the *None* object; this method is used to retrieve transactions, gantries or vehicles based on their unique identifiers;
- *contains* method returns *True* if the collection given as the first argument contains an element with the identifier given as the second argument; otherwise, the method returns *False*;
- *load_gates* method is used to load data about gates from a csv-file, the name of which is given as an argument of the method; based on the information contained

in the file with gate parameters, instances of the *Gate* class representing the appropriate gantries are created; created objects are added to the *gates* collection in the order they were read from the data file, and the items are sorted in ascending order by *id* value;

- *load_transactions* method: allows you to load the information about transactions contained in the file given as the argument to the data model; the method creates *Transaction* objects that are added to the *transactions* collection in the order they were read from the file;
- *set_vehicles* method is designed to create an instance of the *Vehicle* class based on data included in the *transactions* collection; if the *vehicles* collection does not contain an object with the vehicle identifier present in the analysed transaction, a new instance of the *Vehicle* class is created, otherwise (the *vehicles* collection already contains a vehicle with the given identifier) the current transaction is added to the collection of transactions associated with the given vehicle; all transactions assigned to the vehicle are sorted in ascending order according to the time of their occurrence;
- function *get_distances* returns a list containing the travel distance for all vehicles registered in the analysed set of transactions; the distance for one vehicle is calculated as the sum of the distances between the gantries (estimated using the *gps_distance* method), listed in the transactions associated with the vehicle; if the transaction list for the vehicle has only one item, the journey distance is estimated as zero;
- *get_time_series* function is used to determine the unevenness of transactions in time; the argument of this method is a time step that defines the width of the time segments for which the number of registered transactions is estimated (by definition, the time step value is 1 h); the method assumes that all transactions analysed within the data model occur during one day, so for a time step of 1 h, the time intervals are a list of tuples (0..0.1), (1..0.2),..., (23..0.24), associated with the hours of the day being analysed; the function returns a list containing the number of transactions recorded during the appropriate time ranges.

5 Results of the Statistical Analysis

Using the developed software, the following parameters characterizing the data set from the viaTOLL system were analysed:

- the number of vehicles registered in the system for the transactions included in the analysed data set,
- the number of trips resulting from the analysis of sets of transactions associated with registered vehicles; the vehicle is considered as the one that have completed a trip within the viaTOLL system, if it has been registered in at least 2 transactions from the analysed data set,

- the number of long-term trips: a journey is assumed as a long-term trip, if its duration (estimated with the *travel_time* method of the *Vehicle* class) is bigger than 0.15 h,
- the number of trips for each of the vehicle classes,
- the average duration of long-term journeys per vehicle classes and for the entire set of vehicles completed the long-term trips,
- standard deviation of the time of long-term journeys, divided by vehicle classes and for the whole set of vehicles.

As part of the analysis, the vehicle classes were considered based on the classification presented in Table 1.

The elimination of trips with zero distances and the use within the completed analysis of the indicator of the number of long-term trips were made in order to remove inaccuracies in the values of statistical indicators caused by small non-representative values in the analysed sample.

5.1 Descriptive Statistics

Data on transactions registered in the viaTOLL system at the first stage was analysed in terms of the variability of parameters for each of the days of the analysed time period (from 2019/01/01 to 2019/01/13). The basic descriptive characteristics were estimated as the lower and upper limit of the range of perceived values, the span of the value range (the difference between the largest and the smallest values), the average value, and standard deviation. The obtained values of the mentioned characteristics are presented in Table 2.

Based on the average value of the number of transactions and the number of vehicles recorded in the system, it can be concluded that the average number of transactions associated with a vehicle is 22.39 transactions per vehicle. On average, the number of non-zero trips (journeys within the viaTOLL system with the number of transactions bigger than 1) is 92.04% of the total number of registered journeys, while the number of long-term journeys (with the travel time within the system over 0.15 h) is 89.75% of the total number of trips (or 97.51% of the number of non-zero trips). The sample of non-zero journeys is used as input data for the analysis of the random variable of travel distance in the viaTOLL system, while the sample of long journeys—for the analysis of vehicle travel time in the system.

The obtained average values of the number of registered non-zero trips divided according to the vehicle classes allowed us to state that the most numerous group of vehicles completing journeys within the viaTOLL system are the vehicles of class 50 (heavy goods vehicles with GVM and GTM over 12 tons): the share of vehicles in this class is 83.37% of the total number of long-term journeys. The smallest group (only 0.30% on average) are vehicles of class 13 (light vehicles with GVM and GTM equal to or less than 3.5 tons).

Table 2 Descriptive statistics of the analysed indicators

Parameter	Minimum value	Maximum value	Mean value	Standard deviation
Number of transactions [records]	376258	4035078	2683358	1284063
Number of recorded vehicles [veh.]	22256	173793	119826	55084
Number of trips (class 13) [trips]	2741	7100	5182	1542
Number of trips (class 14) [trips]	217	794	589	150
Number of trips (class 30) [trips]	2063	6169	4966	1129
Number of trips (class 41) [trips]	786	10469	6908	3562
Number of trips (class 42) [trips]	197	5722	3438	2267
Number of trips (class 50) [trips]	12953	128781	86458	42142
Total number of trips [trips]	20321	160798	110288	51033
Number of long-term trips [trips]	18957	157491	107540	50549
Mean travel time [h]	3.65	7.72	6.11	1.203
Standard deviation of travel time [h]	4.21	6.58	5.30	0.615

The average travel time for all vehicles in the analysed sample equals 6.11 h. The longest journeys are taken by vehicles of the 30 class (coaches with more than 9 seats): the journey duration for vehicles of this class is greater than the average at around 35%. The shortest journeys are made by class 14 vehicles (light vehicles with a GVM of 3.5 tons or less and capable of towing a trailer): travel times are less than the average at around 41%.

It also should be noted that the duration of long-term journeys is differentiated unevenly in terms of the vehicle class: the average value of the coefficient of variation for the standard deviation of the travel time is 0.113, wherein the travel time of the vehicles of class 30 has the lowest variability of the standard deviation (on average, about 67% below the average value), and the travel time of the vehicles of class 13 has the biggest variability (on average above 47% from the average value).

5.2 Dynamics of Changes in the Analysed Indicators

As the travel parameters within the viaTOLL system vary not only in terms of еpyшк structure, but also in terms of time—for days of the week, as part of the analysis, the dynamics of change during the analysed period was additionally examined for the following indicators:

- average number of transactions noted during a trip,
- average travel time,
- variation coefficient of the travel time,
- share of vehicles for each class of vehicles,
- hourly number of transactions generated in the system.

Apart from the value on the holiday (2019/01/01), the ratio of the average number of transactions per vehicle is slightly varied, as can be seen from the chart in Fig. 3: the deviation from the average value (21.92 transactions/veh.) between −3.38 and 6.14%.

The average time of long-term journeys within the viaTOLL system differs in terms of vehicle class (Fig. 4): the longest journeys are made with vehicles of class 30, and the shortest—usually with vehicles of class 14 (except for holiday days and weekends, when the class 42 vehicles have the lowest values of the mean travel time). The attention should be payed to the evident seasonality of the average time values: the shortest trips, regardless of the vehicle class, are made on holidays and Sundays, while the longest trips are made in the middle of the week—on Wednesday and Thursday. For all vehicle classes in total, the length of journeys on Sundays is 21% lower than the average for the analysed period, and for the weekly peak the journeys are 22% longer than the average.

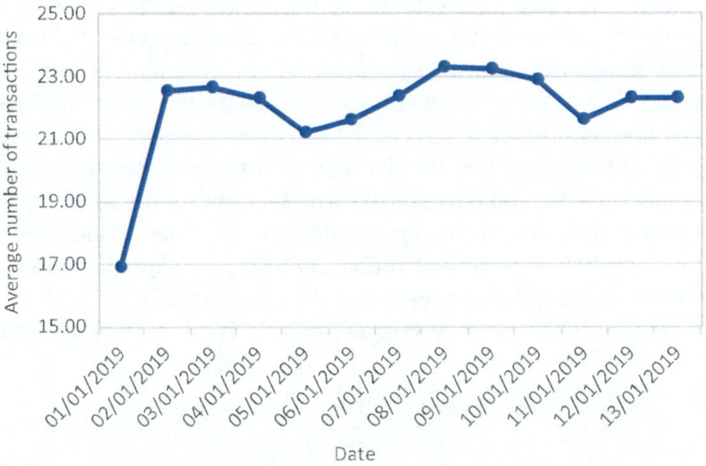

Fig. 3 Dynamics of the average number of transactions recorded for a vehicle

Fig. 4 Dynamics of the average travel time

The dependence of the differences between the time of long-term travel for different days of the week is also clearly visible (see Fig. 5). The coefficient of variation (the ratio of the standard deviation to the expected value) is significantly higher for holiday days and weekends: the value of the coefficient for all vehicle classes in total is bigger on average by 17% for these days.

The study of the dynamics of changes in the number of transactions for hours of the day in the analysed period (see Fig. 6) allows to distinguish patterns of distribution of the number of transactions. The shape of the curve for workdays is similar: from 1:00 to 9:00 a steady increase is perceived, followed by the peak segment from 9:00 to 14:00, after which the number of transactions steadily decreases until the end of the day.

6 Conclusions

The proposed methodology of the data analysis from the traffic toll system allows researchers and traffic engineers to process big amounts of information that is received in real-time. The methods implemented within the core classes library return the results of estimates for the basic parameters of traffic flows, such as the flow

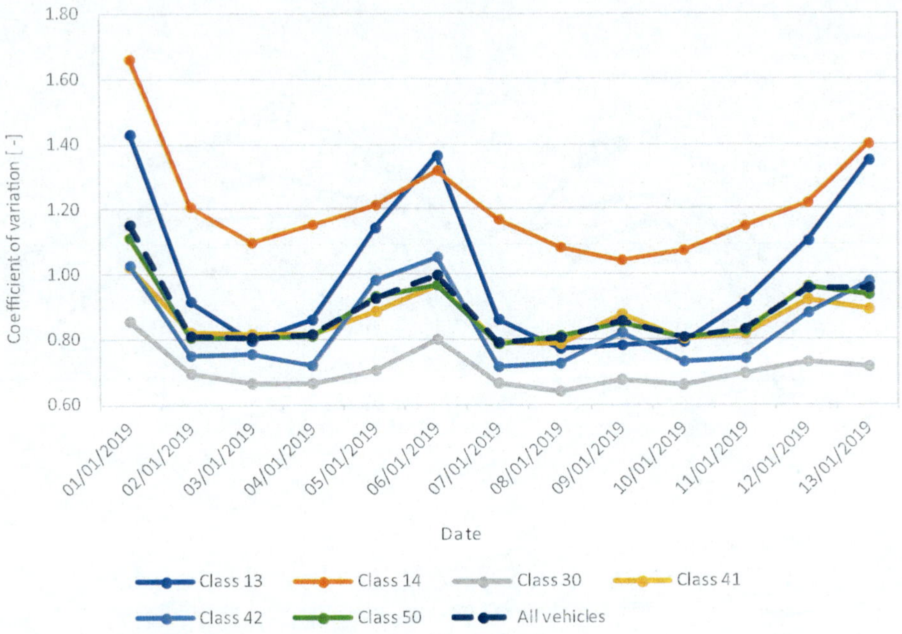

Fig. 5 Dynamics of the variation coefficient for travel time

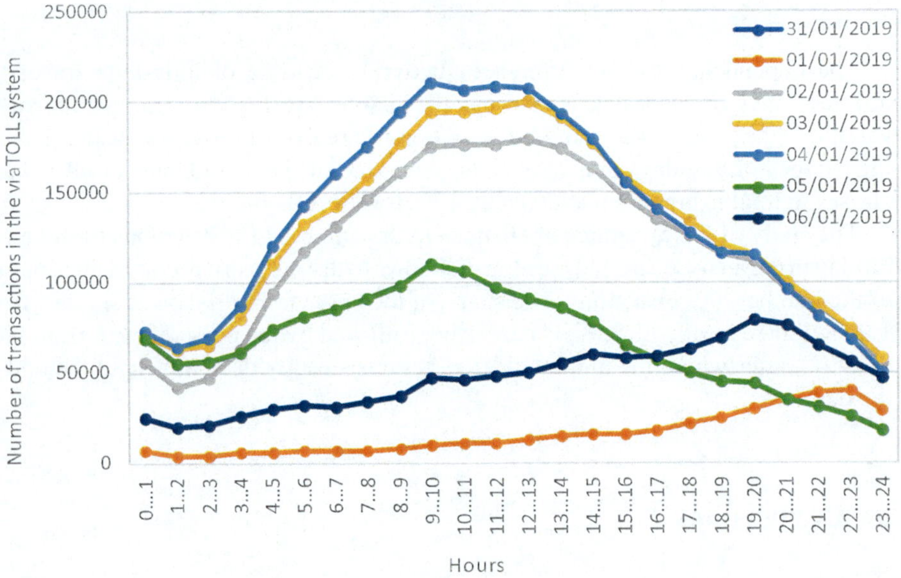

Fig. 6 Dynamics of the number of transactions for the period from 01/01/2019 to 01/06/2019

intensity, average speed, and structure. Furthermore, the proposed approach allows analysing dynamics for the key traffic indicators.

The results obtained based on the case study from the viaTOLL system have the efficiency of the developed methodology and the corresponding software implementation. The evaluated parameters of traffic from the viaTOLL system were used as initial data for the advanced models of the transport system of Poland.

As the further research directions, the development of methods for forecasting of traffic parameter should be mentioned. To integrate the real-time data analysis procedures into the complex transport system model, the algorithms will be developed for including the toll gantries into the transport network model as the vertices of the graph representing the road network.

Acknowledgements Research described in the chapter is prepared in the frame of the project: "Intelligent system for the production of road and maritime transport statistics with the use of large volumes of data to shape the country's transport policy" under the Strategic Program of Scientific Research and Development "Social and economic development of Poland in the conditions of globalizing markets" GOSPOSTRATEG.

References

1. Giuffrè O, Granà A, Tumminello ML (2017) Exploring the uncertainty in capacity estimation at roundabouts. Eur Transp Res Rev 9, Article number: 18. https://doi.org/10.1007/s12544-017-0238-8
2. Granà A, Giuffrè T, Macioszek E, Acuto F (2020) Estimation of passenger car equivalents for two-lane and turbo roundabouts using AIMSUN. Front Built Environ 6:86. https://doi.org/10.3389/fbuil.2020.00086
3. Sierpiński G, Macioszek E (2020) Equalising the levels of electromobility implementation in cities. In: Mikulski J (ed) Research and the future of telematics, communications in computer and information science, vol 1289. Springer, Heidelberg, pp 165–176
4. Macioszek E, Sierpiński G (2020) Charging stations for electric vehicles—Current situation in Poland. In: Mikulski J (ed) Research and the future of telematics, communications in computer and information science, vol 1289. Springer, Heidelberg, pp 124–137
5. Angarita-Zapata JS, Masegosa AD, Triguero I (2019) A taxonomy of traffic forecasting regression problems from a supervised learning perspective. IEEE Access 7:68185–68205
6. Lopes J, Bento J, Huang E, Antoniou C, Ben-Akiva M (2010) Traffic and mobility data collection for real-time applications. 13th International IEEE conference on intelligent transportation systems. IEEE, Funchal, pp 216–223
7. Kim JY, Karauchi F, Uno N, Hagihara T, Daito T (2014) Using electronic toll collection data to understand traffic demand. J Intell Transp Syst 18(2):190–203
8. Soriguera MF (2016) Highway travel time measurements from toll ticket data. In: Highway travel time estimation with data fusion. Springer tracts on transportation and traffic, vol 11, pp 109–155
9. Lentzakis AF, Seshadri R, Akkinepally A, Vu VA, Ben-Akiva M (2020) Hierarchical density-based clustering methods for tolling zone definition and their impact on distance-based toll optimization. Transp Res Part C: Emerg Technol 118, art. 102685
10. Soorya VB, Arkatkar SS, Vanajakshi L (2020) Exploratory study on approaches for traffic count prediction; using toll-way traffic count. Transp Res Procedia 48:3880–3895

11. Díez-Gutiérrez M, Andersen SN, Nilsen ØL, Tørset T (2019) Generated and induced traffic demand: empirical evidence from a fixed link toll removal in Norway. Case Stud Transp Policy 7(1):57–63
12. Zhao N, Qi T, Yu L, Zhang J, Jiang P (2014) A practical method for estimating traffic flow characteristic parameters of tolled expressway using toll data. Procedia—Soc Behav Sci 138:632–640
13. Neuhold R, Garolla F, Sidla O, Fellendorf M (2019) Predicting and optimizing traffic flow at toll plazas. Transp Res Procedia 37:330–337
14. Karsaman RH, Mahendra Y, Rahman H, Sulaksono S (2014) Measuring the capacity and transaction time of cash and electronic toll collection systems. J Eng Technol Sci 46(2):180–194
15. Bains MS, Arkatkar SS, Anbumani KS, Subramaniam S (2017) Optimizing and modeling tollway operations using microsimulation: case study Sanand Toll Plaza, Ahmedabad, Gujarat, India. J Transp Res Board 2615(1):43–54
16. Puts MJH, Daas PJH, Tennekes M, Blois C (2019) Using huge amounts of road sensor data for official statistics. AIMS Math 4(1):12–25
17. Michau G, Nantes A, Bhaskar A, Chung E, Abry P, Borgnat P (2017) Bluetooth data in an urban context: retrieving vehicle trajectories. IEEE Trans Intell Transp Syst 18(9):2377–2386
18. Janecek A, Valerio D, Hummel KA, Ricciato F, Hlavacs H (2015) The Cellular network as a sensor: from mobile phone data to real-time road traffic monitoring. IEEE Trans Intell Transp Syst 16(5):2551–2572
19. Burnos P, Gajda J, Piwowar P, Sroka R, Stencel M, Zeglen T (2007) Measurements of road traffic parameters using inductive loops and piezoelectric sensors. Metrol Measurement Syst 14(2):187–201

Optimizing Last Mile Delivering Through the Analysis of Shoppers' Behaviour

Olga Kunytska, Antonio Comi, Viktor Danchuk, Kateryna Vakulenko, and Serhii Yanishevskyi

1 Introduction

E-commerce is growing exponentially, accounting for about 10% of the global retail trade and driving the sector's growth. Global e-commerce revenues, which were less than $ 20 billion in 1997, are estimated by The Fashion and Apparel Industry to grow up to $ 712.9 billion by 2022 [1]. The main reasons for this phenomenal growth are the expansion of online access through smartphones and other portable devices, innovative sales technologies (e.g. artificial intelligence, chat bots, multi-channel retail, hardware innovation, social media marketing), the development of new sales markets outside the West, ease and speed of receiving the purchased goods.

In Europe, online shoppers have increased by 85% over the past decade (Consumer Condition Scoreboard 2017: 92) and e-commerce is expected to account for 36% of

O. Kunytska (✉) · V. Danchuk · S. Yanishevskyi
National Transport University, Mykhaila Omelianovycha—Pavlenka Street 1, 01010 Kiev, Ukraine
e-mail: olga.kunytska@ntu.edu.ua

V. Danchuk
e-mail: vdanchuk@ukr.net

S. Yanishevskyi
e-mail: s.yanishevskiy@ukr.net

A. Comi
Department of Enterprise Engineering, University of Rome Tor Vergata, Via Del Politecnico 1, 00133 Rome, Italy
e-mail: comi@ing.uniroma2.it

K. Vakulenko
O.M. Beketov National University of Urban Economy in Kharkiv, Bazhanova Street 18, 61002 Kharkiv, Ukraine
e-mail: vakulenko.e@ukr.net

G. Sierpiński and E. Macioszek (eds.), *Decision Support Methods in Modern Transportation Systems and Networks*, Lecture Notes in Networks and Systems 208,
https://doi.org/10.1007/978-3-030-71771-1_9

global trade by 2030 (World Economic Forum). In 2019, the global e-Commerce market grew by 17.9% compared to the previous period and amounted to $ 3.46 trillion. China and the United States are the world leaders in the e-commerce market: they account for about 40% of the market [2].

According to a study by EVO (Prom.ua, Shafa.ua, Bigl.ua, Crafta.ua, Izi.ua) and OLX service, in 2019 e-commerce turnover in Ukraine amounted to 76 billion UAH. (€ 2.3 billion). This is 17% more than in the same period in 2018. On average, consumers spent € 6 – € 30 per purchase.

E-commerce is the process of purchasing goods and services over the Internet [3]. It is the result of the popularization of transactions on the Internet, and hence the possibility of concluding sales contracts in a virtual environment. Some of the most important benefits of e-commerce from a shopper's point of view are [4, 5]:

- Availability 24/7,
- Home delivery/no need to go to the store,
- Easy comparison of offers,
- Attractive prices,
- Easy search for rare/specialized products,
- A significant amount of information about products,
- Wider assortment than in traditional stores,
- The right to return any purchases made "remotely".

This new shopping trend can have impact on freight traffic in urban areas because products purchased have to be delivered to customer at home through delivery tours that cannot always be optimised and cause growth of traffic impacts [6]. Besides, additional costs for repeated deliveries can occur. In fact, attended home deliveries require customers and logistics service providers to agree on a service time window in order to avoid failure of delivery. For logistics service providers, accepting more deliveries can yield more revenues, but it can make it harder to reliably make deliveries within the service time windows.

From an end consumer's point of view, on-line retail is associated with a number of benefits such as greater product choice, the ability to obtain goods not sold locally, better price comparison, etc. From a logistics point of view, however, the solution for delivery is very demanding. Efficient and reliable logistics are a key factor for the economic success of on-line shops, and shipping costs are one of the biggest concerns for on-line customers. Especially the *not-at-home problem* has to be treated, which results from the delivery of goods requiring the presence of the customer. This leads to complex planning problems within the last leg in supply chains, i.e., the last mile to the consumer.

By doing so, e-commerce is fundamentally changing the way we do business internationally, nationally, domestically, and locally. In the transportation sector, online shopping is expected to lead to new approaches to determining the distribution of goods in cities, as well as to change the behavior of citizens when traveling [7].

This can affect transportation directly as transport organizations introduce new methods and businesses and consumers change their e-commerce delivery requests. The scale and speed of change increase the relevance of planning infrastructure and services in response to changing consumer needs. Rationally organized delivery to the final consumer is often decisive and determines the choice of an online store. Hence, the key question for e-commerce is the last mile delivery (LMD). Although home delivery is convenient for the customer, LMD poses significant logistical challenges for companies, including scheduling and routing problems [5].

However, new opportunities are emerging for developing profitable LMD strategies [8]. Specifically, companies can influence customer behavior by choosing suggested dates or time slots (capacity controls) and associated fees (pricing controls). These solutions are ultimately aimed at balancing capacity utilization and increasing the profitability of delivery operations. However, these decisions are subject to the consent of the end consumer to use them.

The purpose of this chapter is to investigate, based on a study of e-commerce consumers, the factors that impact on way of purchasing (i.e. on line or at stor) as well as the factors influencing the choice on how to receive the e-purchases (e.g. at home or at pick up points/locker points).

2 Literature Review

The analysis shows that the study of transportation problems in the case of e-commerce is versatile. Nowadays, there are developed theoretical approaches, models, and methods for solving this problem. Moreover, existing studies do not allow to fully draw conclusions about the degree of relationship between e-commerce and transport, the particularities of the delivery of goods under such conditions. There is a general assumption that e-commerce is a growing aspect of the increasingly digital era and e-commerce continues to grow and expand into new markets and industries [9].

In [10] envisage an increase in e-commerce in older age groups, an increase in online shopping habits among younger age groups, a decrease in the number of physical stores, an increase in the number of e-commerce domains (i.e. groceries) and increased use of smartphones. There is a growing amount of research on the impact of e-commerce on freight transport.

A key feature of e-commerce is home delivery. The increase in online shopping has led to an increase in truck travel by courier services. Generally speaking, e-commerce enhances the trend towards an increase in parcel volume with a simultaneous decrease in their weight and, as a result, an increase in travel frequency [11].

The growth of urban freight transport due to changes in the supply chain (for example, on-time delivery and home delivery) leads to an increase in supplies and an increase in the number of freight vehicles in residential areas [12]. There are forecasts that with an increase in the number of online orders, an increase in the number of commercial vehicles in residential areas, as well as an increase in congestion, air pollution and noise problems (environmental issues), and an increase in logistics costs, and therefore product prices (economic issues).

Several studies indicate that e-commerce has the potential to enhance personal travel as it replaces travel to physical stores [13–15], but this effect can be overestimated as studies tend to ignore chain travel [16].

Some studies have investigated how end consumers make the choice between e-shopping and at-store shopping, showing that decision making is influenced both by socio-economic characteristics (age, income, level of education; [17–23], built environment/land use [24, 25], as well as time resource [26, 27].

In [28] the joint choice of pre-purchase and purchase shopping way (i.e. in store and on line) are analysed. The analysis showed that the combinations of "only-store pre-purchase plus store purchase" and "only internet pre-purchase plus internet purchase" occur substantially more often than independent choices.

Nuzzolo and Comi, 2016 [29] presents a new model system for modelling shopping choices, including e-shopping, based on an analysis of demographic and socio-economic factors that influence the way the final consumer purchases and the creation of next trips.

The relationship between e-commerce and travel behaviour is also examined in [23], where it is determined that the impact of e-commerce on personal travel is highly dependent on the consumer, household and location.

Research by [30] defines a typology of e-commerce consumers based on online shopping motivation. Analyzes of these motives include the convenience of online shopping, the physical orientation of the store (i.e., the ability to instantaneously own a product and have social contact), the use of information in planning and shopping, and the search for diversity in the context of online shopping.

Repetitive deliveries of low cost goods are typical for e-commerce, and in practice, lead to longer vehicle routes. Moreover, during the standard operating hours of suppliers, customers are often far from their home. This creates the need for repeated delivery to the same address, and in some cases, even a third courier visit is required. The problem of re-delivery affects 20-30% of recipients [31]. At the same time, the mileage of the logistics operator's vehicles increases and large emissions of pollutants occur.

According to [32], the last mile of delivery is considered one of the most expensive, most efficient and most polluting parts of the entire supply chain. In this regard, solutions are being created to minimize the negative consequences of the growth of e-commerce. One solution is to use the parcel lockers. The logic of their work is based on the assumption that it is the recipient who comes for the package at a certain point at a convenient time for him. This practice leads to a reduction in the courier's vehicle mileage, which leaves many parcels in one place. For the supplier, this translates into a lower cost for each delivery. The recipient, in turn, is not limited by the courier's schedule. The parcel lockers system reinforces the global last-mile experience. According to a recent analysis of the global parcel lockers market by the Belgian company International Post Corporation, parcel lockers are already widely used in Finland, Denmark and China with up to 43% of usage. Consumers prefer the parcel lockers method, citing 24/7 availability, choice of delivery location and lower shipping prices.

Therefore, although home deliveries are usually preferred by on-line shoppers [33], from a planning point of view, there is a rising interest in deloping alternatives that satisfy both consumer demand and transport operators to optimize parcel distribution through consolidated shipments. In Europe, pick-up points (i.e. stores providing parcel drop-off and pick up services) are fast-growing solutions. The largest pick-up point network is the Packstation network operated by DHL/Deutsche Post in Germany (2500 locations around the country; [34]).

According that in the next future the e-purchases will increase and the impacts due to home deliveries could raise, the chapter investigate the behavioral characteristics of the end consumer in choosing the place of receiving the parcel, points out the factors that influences the choice of the user of the parcel lockers among other alternatives available.

3 Research Method and Experimental Data

3.1 Data Collection Procedure

This work is based on a survey among consumers from different regions of Ukraine, in which the behavior of consumers in their choice of the method of purchase and the choice of a place of delivery was studied.

The purpose of the survey was to study the dependencies among the demographic and economic data of consumers and their behavior in choosing the method of making a purchase (online or at-store) and choosing the method of receiving the parcels, which is a factor influencing the solution of the last mile of delivery. The study focused on comparing different ways of getting online shopping and investigated the factors influencing consumer choice.

The survey was carred out online using existing google applications for a month from mid-August to mid-September 2020. The volume of answers was recruited in order to achieve a sample representative set in terms of the number of the population of the region. Determination of sample size (1), (Table 1).

$$n = \frac{Z^2 pq}{\Delta^2} \tag{1}$$

n—sample size,

Z—is a coefficient that depends on the confidence level chosen by the researcher. In the study, the confidence level is 95%, at which Z equal 1.96;

p—the proportion of respondents with the presence of the studied feature (equal to 0.5);

q = 1—p is the proportion of respondents who do not have the studied feature (equal to 0.5);

Δ—confidence interval error (equal to 5%).

Table 1 Determining the required sample size

Characteristics	The values
Confidence probability, confidence, reliability	95%
Confidence interval error	± 5%
General set of respondents*	7,000,000
The required sample size	384
The number of answers received	537
Rejected as incorrectly filled	3

*The value of the Ukrainian audience of online buyers, determined by ubr.ua on the basis of statistical data of online services OLX and research company GfK, was accepted as the general set of respondents

Thus, the number of responses accepted for analysis is 534, under the age of 22 and older than 56 years, the distribution of which according to the relevant parameters is described in detail in Table 2.

3.2 *Research the Typology of Online Shoppers Based on Purchase Motivation*

During the survey, some of the questions were formulated similarly to the study of [30] on the choice of method of purchase and determine the motives that influence this choice (Table 3). This allowed to carry out an analysis to confirm the results of the work of [30], where on the basis of the obtained data on the motivation of online purchases, an appropriate typology of buyers was developed. It should be noted that there are currently few works that study typologies in the context of the Internet, although many traditional shopping typologies have been developed, mainly based on customer visits to certain stores.

In [30] study examined the motives for shopping that are important to the online context, including convenience, information retrieval, immediate ownership, social interaction, retail shopping experience, and diversity seeking. Accordingly, consumers, based on purchases, are divided:

"Convenient buyer"—motivated by the prospects of general convenience of online shopping. This segment also shows less focus on the physical store (for example, less motivated by the prospect of immediate possession of purchased goods or services and social interaction), as well as less desire for diversity in retail channels. In the case of this study, the choice of the most important factor on the influence of the choice of method/place of purchase, chose the opportunity to save time and effort.

"Diversity Seeker"—moderately motivated by the convenience of shopping online, but much more—the search for diversity among retail alternatives, product types and brands. The Diversity Seeker shows close to average scores on physical store orientation, as well as a propensity to plan shopping and shopping trips. In this

Table 2 Description of the initial data used in the analysis (own research)

Age	N	%	Gender	N	%
Less than 22	134	25	Male	221	41
22-35	183	34	Female	313	59
36-45	165	31	**Income per month (total, all family members)**		
46-56	29	5	Less than €160*	11	2
56 years and over	24	5	€160–€320	61	11
Education			€320–€1000	264	50
Less than high school graduate	62	12	€1000–€2000	148	28
Professional (college and technical)	36	7	€2000 +	50	9
High school graduate	377	70	**Area of residency (by population)**		
Scientific	59	11	The most important cities (over 1 million)	379	70
Does not have	0	0	The biggest cities (from 500 thousand to 1 million)	19	4
Household size (including all adults and children)			Large cities (from 250 to 500 thousand)	33	6
1 person	52	10	Medium cities (from 50 to 250 thousand)	37	7
2 persons	112	21	Small towns, including urban-type settlements (up to 50 thousand)	47	9
3 persons	180	34	Rural settlements	19	4
			Main activity		
4-5 persons	173	32	Employed	375	70
			Student	104	20
6 or more persons	17	3	Pensioner	7	1
			Not working	48	9

[a]Minimum wage in Ukraine on 09/01/2020, source—minfin.com.ua

study, this group includes consumers who determined the possibility in the purchase process to change the store, brand, product and those who chose the opportunity to obtain detailed information about the product, to compare with others.

"Balanced shopper"—this type of motive is similar to the motivation of the seeker of diversity in his or her desire for convenience and the least in its propensity to plan purchases or search for information. A balanced customer is moderately motivated by the desire for diversity and demonstrates close to average scores in terms of physical orientation of the store. In the study, this group includes consumers who chose previous shopping experience as the main criterion.

Table 3 Comparison of the distribution of typologies of purchases in the context of the Internet in this study and in the study of Andrew J. Rohm

Online shopping type	% of consumers in group in their own study	Rank	% of consumers in this group in a study by Andrew J. Rohm	Rank	The difference between the results of the study in relation to the study of Andrew J. Rohm,%
Convenience shopper	23	2	11	4	+209
Variety seeker	50	1	41	1	+21
Balanced buyer	11	4	33	2	-67
Store-oriented shopper	16	3	15	3	+6

"Shop-oriented customer"—is characterized by the lowest level of desire for convenience when shopping online. This type of customers appreciate the physical orientation of the store, i.e. the ability to immediately own the goods and the possibility of social interaction. According to the results of the study conducted in this study, this group of consumers included those who, when choosing the most important factor in the influence of the choice of method/place of purchase, chose the opportunity to receive the product immediately after purchase and the ability to communicate with the seller or other buyers.

A comparison of the results obtained in this work and in the work [11] indicates a change in the priorities of consumers when choosing a method/place of purchase. In the first place also remains a group of "Diversity Seekers", who value most of all the opportunity to obtain information about the product and compare with others. In second place there is the group "Convenient Buyer", which previously occupied the 4th place. Thus, the ability to save time and effort, which is a key factor for this type of consumer, has become more important for buyers. The motive of previous experience, which is decisive for the group "Balanced Buyer" decreased by 67%, and the volume of shoppers focused on the store, who seek to instantly receive the goods and have social interaction at the time of purchase remained in the same position.

3.3 Impact of Pandemic COVID 19 on Online Shopping Behavior

The growing impact of the motive of saving time and effort, which is most characteristic of online shoppers, can be explained by the results of quarantine for the declared pandemic COVID 19. The study compared the experience of online shopping before quarantine because of COVID19 and after quarantine (Table 4).

Table 4 Comparison of the experience of online shopping before the introduction of quarantine because of COVID19 and after the start of quarantine (own study)

	%		%	%
Changing the frequency of Internet orders with the introduction of quarantine		Frequency of orders	Before COVID19	With the beginning of COVID19
Order more often	33	once or more a week	6	11
Order less often	4	2–4 times a month	22	28
The frequency of orders has not changed	63	1–2 times a month	28	27
Changing the cost of the Internet—orders after the introduction of quarantine		1–2 times during 3 months (quarter)	32	23
Increased by 2 times or more	9	1–3 times a year	10	9
Increased up to 2 times	15	less than once a year	2	2
They remained the same	68	**Approximate cost of online ordering for the last year**		
Decreased to 2 times	5	Up to 30 €	10	
Decreased by 2 or more times	3	From 30 € to 300 €	44	
		From 300 € to 600 €	26	
		From 600 € to 1 500 €	12	
		More than 1 500 €	8	

The study determined the nature of consumer orders (Table 5) when making an online purchase.

Table 5 Characteristics of orders (own research)

Order method	%	Type of orders	%
		Only personal	65
Choose home delivery	32	More often personal than working	21
Take away the goods yourself	68	Equivalent orders—both personal and work orders	12
		More often working than personal	1
		Only working	1

The key logistical problem of e-commerce arises at the last stage of delivery and is known as the last mile problem. The geographical dispersion of recipients necessitates the organization of deliveries in almost every place of residence of a potential buyer. For the logistics operator, this is due to the optimization of the place, mode and time of receipt of the parcel, delivery time, the average cost of delivery, envirommental impact etc. (e.g. [35–39]).

Therefore, solutions are sought to optimize the last stage of order delivery to the consumer. One solution is to use collection points. The logic of their work is based on the assumption that the recipient comes for the parcel at a certain point: a store, kiosk or parcel locker. This practice reduces the mileage of the courier car, which leaves many parcels in one place. For the supplier, this reduces the cost of each delivery. The recipient, in turn, is not limited by the courier's work schedule.

4 Results of the Study and Duscussion

4.1 Choice of Delivery Method by on-Line Buyers

Parcel lockers (post machines) as a way to solve the above identified problems of the last mile of delivery is an interesting and innovative solution for any type of city, beneficial for both customers and online stores. Given the growing number of B2C e-commerce customers, this delivery method meets the needs of today and future society—both in terms of environmental friendliness and impact on the congestion of the urban transport network, and in terms of convenience for consumers.

The most important goal of the introduction of parcel lockers is to reduce the number of deliveries in the urban area, including unsuccessful courier deliveries and the subsequent return of goods by couriers and postal services. This helps to reduce unnecessary vehicle mileage and reduce the impact of these trips on city congestion.

The most important factor in the effective use of parcel lockers is their correct location within the city. According to the data obtained by Stanisław Iwan et al., the most effective locations of parcel lockers are places associated with the presence of:

- local popular spots, as an example, next to round-the-clock shops (high population density living in the area),
- pedestrian zone with heavy traffic in city centers,
- proximity to shopping centers and supermarket parking lots,
- proximity to transport hubs—bus/metro stations
- gas station,
- auto repair stations,
- business centers.

The importance of the convenience factor is confirmed by the results of this study (Table 6), in which 62% of respondents determine the convenience factor (the ability

Table 6 Assessment of the importance of factors when choosing a method of delivery

Choosing the most influential factor when choosing a delivery method	%
Home/work delivery, so as not to waste time picking up the order	18
Ability to pick up the goods at a convenient time, even if you need to intentionally go to the place of issue	62
Opportunity to save on delivery	14
No need to wait in line	4
No need to contact the operator	1
Other	1

to pick up the parcel at a convenient time, even if you need additional travel to the place of delivery)—as a major factor in choosing a method of delivery.

To assess the probability of choosing a parcel locker as a method of delivery of Internet purchases and the analysis of the reasons leading to such a choice, the following issues were investigated:

- under what conditions the recipient will prefer parcel lockers in comparison with other methods of receiving the order;
- whether the recipient's income affects the choice of delivery method;
- whether the availability of the vehicle increases the probability of choosing a parcel locker;
- whether there is a dependence of the method of delivery, which the consumer chooses with the distance, he is willing to cover to the place of the delivery of the parcel and whether the presence of his own car affects this choice.

The first step in answering these questions was to evaluate the comparison of existing delivery experience and the delivery method most often chosen by the consumer (Table 7).

As can be seen from the results, the experience of using parcel lockers has only a small number of consumers. Therefore, in the study for clarification, an assessment of consumer awareness of this method of delivery was conducted (Fig. 1).

Table 7 Choice of the delivery method

Delivery method, which has experience of use, %		The most common delivery method, %	
Courier delivery	65	Courier delivery	16
Delivery point	96	Delivery point	80
Mobile hotspots	21	Mobile hotspots	3
Parcel lockers	20	Parcel lockers	1

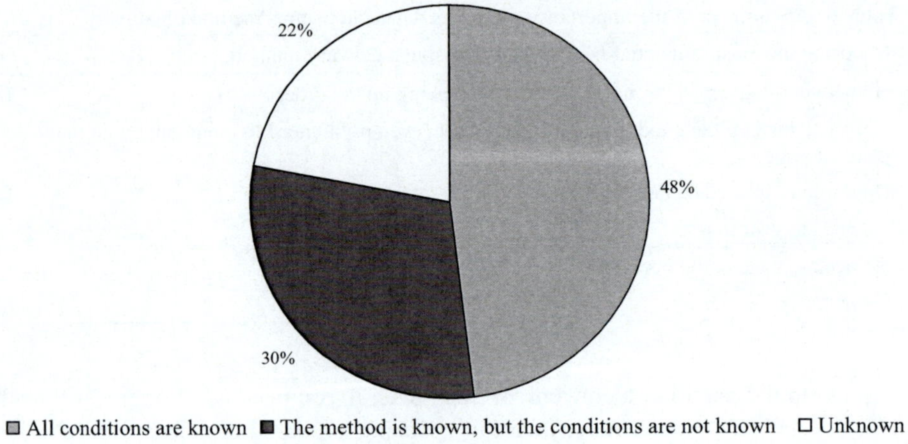

Fig. 1 Assessment of the awareness of the delivery of Internet orders through the parcel lockers

4.2 Comparative Analysis of Factors Affecting the Choice of Delivery Method for Online Orders

It can be concluded that the unpopularity of delivery by post is due to low awareness of the public about such a service—more than 50% of respondents do not know the details and conditions of delivery in this way.

This confirms the comparison of the most popular type of delivery—to the point of delivery located behind the parcel locker. For this purpose, during the survey, respondents were provided with a photo of the parcel locker and an explanation of the conditions of use. Then comparisons were made under the following conditions (Fig. 2):

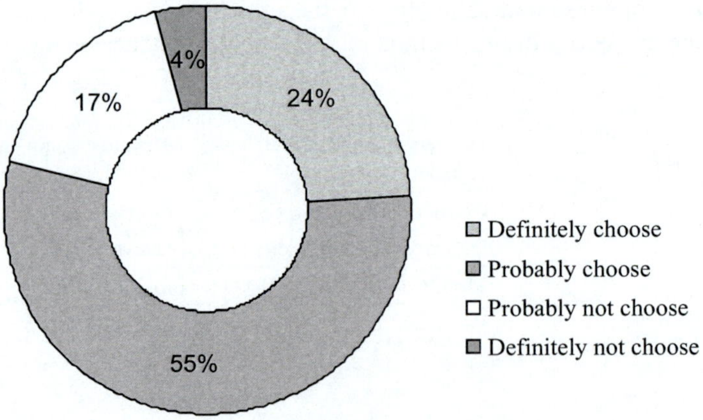

Fig. 2 The results of comparing the parcel locker s with other methods of delivery

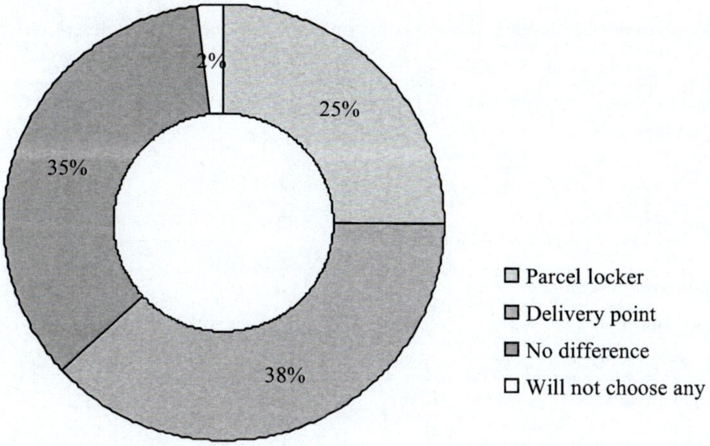

Fig. 3 The results of comparing the parcel locker with the point of delivery with other conditions being equal

- If delivery to the parcel locker will have the lowest price and the parcel locker will be located near the metro station or near work/home, how likely is it that you will choose this method of delivery?
- If delivery to the point of delivery and delivery to the parcel locker are equal in distance and the same cost, which method of delivery will you choose? (Fig. 3)

The result of the comparison determines the general interest in the use of parcel lockers. Thus, the answer to the first question about the conditions under which the recipient will prefer parcel lockers in comparison with other methods of receiving the order, is in the creation of the conditions for easy approachability and affordability. However, under any of the proposed conditions, parcel lockers chaines lose to the main method chosen by customers—delivery points. One of the main factors of advantage of such delivery points is the possibility to see the product before receiving and paying. This feature is not available in parcel lockers.

The next problem investigated in the work is the influence of the recipient's income on the choice of the delivery method.

The chapter calculates the dependence of income on the choice of delivery of the Internet orders (Table 8), which does not give clear conclusions about the use of parcel lockers, due to the small percentage of their use. However, there is a tendency to increase the percentage of choice of courier delivery with increased revenue.

The choice of delivery method depending on the availability of the vehicle is estimated (Fig. 4).

In general, 49% of respondents to this study have their own car, and 51% do not. It is determined whether this factor influences the choice of online order delivery method chosen by the consumer.

The presence of the car does not significantly affect the choice of delivery method and the distribution between the options remains almost equal. However, there is an increase in the percentage of use of methods that involve personal obtaining of

Table 8 Estimation of influence of the income on a choice of a way of delivery of Internet orders

	Income				
The method of delivery that the respondent prefers	Less than €160	€160–€320	€320–€1000	€1000–€2000	€2000+
At the point of delivery	73%	77%	80%	70%	72%
Courier delivery	9%	2%	10%	21%	18%
Mobile hotspots	0%	3%	2%	5%	6%
Parcel locker	0%	3%	0%	1%	2%
Did not answer	18%	15%	8%	3%	2%

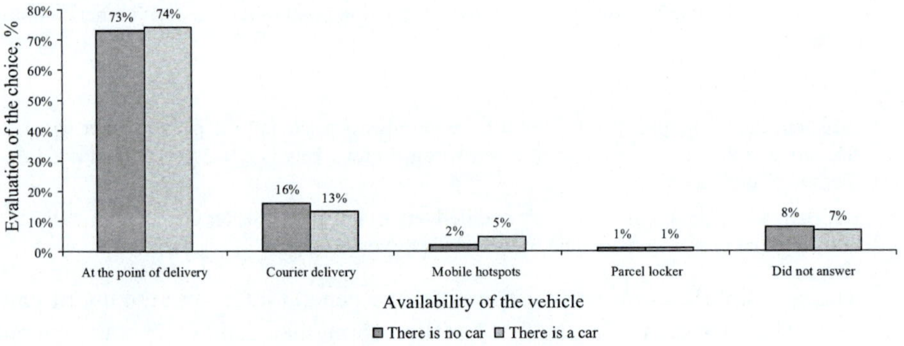

Fig. 4 Evaluation of the choice of delivery method depending on the availability of the vehicle

goods—at the point of delivery or mobile points with the presence of a car (4% in total), and with the absence of own vehicle—an increase in courier delivery by 3%. Whether this leads to an increase in the number of trips was not determined by the study. At the same time, the probability of choosing a parcel locker is equal in both categories.

The ratio of the distance that the consumer is ready to cover to the place of the receipt of the parcel with the delivery method he chooses and whether the presence of his own car affects this choice is determined (Table 9). The results show that the presence of the vehicle does not affect the distance that the recipient is ready to cover to the point of issue. The distance also does not give clear dependencies on the choice of delivery method.

Indirectly, it can be concluded that receiving online orders by consumers personally does not create their additional trips, because in most cases (61%) they choose the option within walking distance. The assessment of the impact on the change in the load of the transport network, depending on the method of receiving the cargo, must be performed from the point of view of the supplier, which may be the next stage of the study.

Table 9 The ratio of the method of delivery, which the consumer chooses taking into account the distance he is willing to cover to the place of receipt of the parcel and taking into account the presence of his own car, %

The method of delivery that the respondent prefers	The approximate distance that the consumer usually covers to obtain an online order							
	Up to 1 km		*1–3 km*		*More than 3 km*		*Always courier delivery*	
	Car	*No car*	*Car*	*No car*	*Car*	*No car*	*Car*	*No car*
At the point of delivery	20	21	10	11	2	2	1	1
Courier delivery	2	2	1	1	1	1	2	4
Mobile hotspots	1	1	1	1	1	1	0	0
Parcel locker	1	1	1	0	1	1	0	0
Did not answer	7							

Figure 5 refines the result obtained by determining the advantages in distances that the consumer covers to receive an order only among those who have their own vehicle. This takes into account the frequency of the vehicle use. The result of the analysis confirms that the presence of the vehicle does not increase the distance of the ordering points. Most often, recipients choose a place of reception within walking distance. Thus, it can be indirectly concluded that the presence of your own vehicle does not create additional trips when receiving online orders.

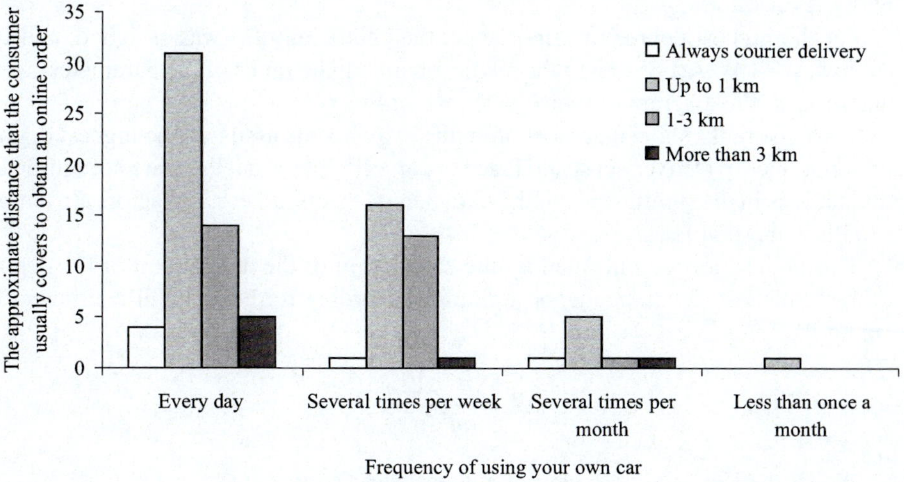

Fig. 5 The ratio of the frequency of use of own vehicles and the distance covered to receive the parcel

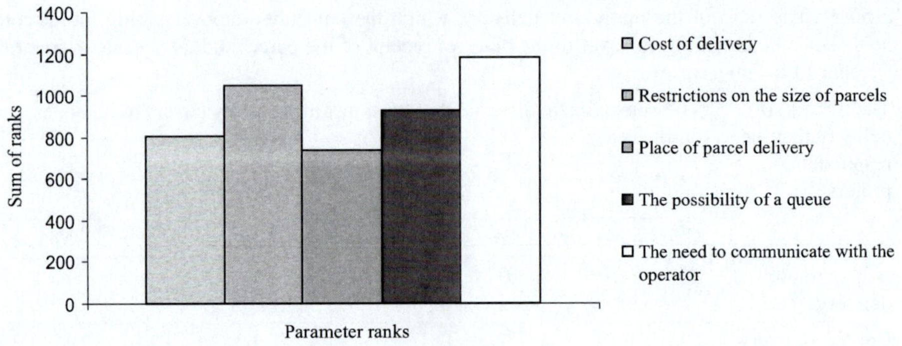

Fig. 6 Parameter ranks when choosing a delivery method

4.3 *The Assessment of the Degree of Agreement of the Respondents' Opinions*

All the results show the advantage of delivery points over other ways to get online orders. Since one of the authors' goals was to assess the choice and prospects for using the Parcel lockers, the study determined the influence of factors on the choice of the method of delivery.

Using the Likert scale, the respondents rated the importance of the parameters when choosing a delivery method, namely: Delivery cost; Restrictions on the size of parcels; Place of parcel delivery; The possibility of a queue; The need to communicate with the operator.

For the highest degree of importance, the coefficient "1" was assigned, and for the lowest—"4". According to the results obtained, the ranks of the parameters were determined, the diagram of which is shown in Fig. 6.

Thus, the ranks show that most often the respondents assigned the highest degree of importance to Delivery cost and Place of parcel delivery, but there are no significant fluctuations in the significance of the parameters, that is, all certain factors are almost equally important for users.

This conclusion is confirmed by the calculation of the assessment of the degree of agreement of the respondents' opinions according to the Kendall coefficient of concordance:

$$W = \frac{12S}{m^2(n^3 - n)} \tag{2}$$

W—is the Kendall coefficient of concordance;
m—is the number of respondents (equal to 534);
n—is the number of factors (equal to 5);
S—is the sum of the squares of the deviation (equal to 129 247).

The value of the coefficient of concordance W for the study is 0.05, which indicates the absence of a single choice among all respondents and the significance of a specific factor that influences the choice of respondents.

5 Conclusions

The growth of the B2C e-commerce market determines the importance of last mile delivery in cities. This changes the volume and characteristics of traffic flows, which can affect traffic and congestion problems. The use of alternative delivery methods, such as parcel lockers, can reduce the number of deliveries. This requires studying the motives of consumers for the choice of delivery methods and studying the influence of consumers on the creation of additional trips related to the receipt of on-line purchases.

The answers of 534 respondents collected in the work showed several trends. Thus, it is indirectly established that the presence of your own vehicle does not lead to covering longer distances to receive the parcel compared to those recipients who do not have their own car. Thus, the presence of a car does not lead to the creation of additional trips by individuals—by car owners when making online purchases. This confirms the comparison of the frequency of use of their own vehicle and the distance that the consumer travels to receive the parcel. Among the consumers who most often use the vehicle, the distances to receive parcels are generally less than 1 km, which are covered by the consumer without the use of a car.

Due to the lack of statistical data, the chapter did not obtain significant results in determining the trends in the selection of parcel lockers as a way to obtain an order/purchase. The reasons for this are not so much the insufficient number of respondents as the low awareness of respondents about this method of receiving parcels—more than 50% of respondents do not know the details and conditions of delivery by this method. However, after getting acquainted with the conditions, and when comparing this method with others, interest in Parcel lockers increases to 55%, given the easy approachability and affordability.

However, it is important to emphasize that this study is based on the reports of only persons receiving the goods. The next step should include data on the activities of transport companies to study how consumer' motivations for the choice of method of delivery affect the creation of additional trips and solve the problems of the last mile of delivery.

References

1. Xenia Dorna, 5 trends in fashionable electronic commerce in 2020. http://www.marketer.ua
2. Snoeck A, Merchan D, Winkenbach M (2020) Revenue management in last-mile delivery: state-of-the-art and future research directions. Transp Res Procedia 46:109–116

3. Turban E, King D, Lee J, Viehland D (2002) Electronic commerce: a managerial perspective, vol 4. Prentice Hall: ISBN 0, 13(975285)
4. Chiu CM, Wang ET, Fang YH, Huang HY (2014) Understanding customers' repeat purchase intentions in B2C e-commerce: the roles of utilitarian value, hedonic value and perceived risk. Inf Syst J 24(1):85–114
5. Moroz M, Polkowski Z (2016) The last mile issue and urban logistics: choosing parcel machines in the context of the ecological attitudes of the Y generation consumers purchasing online. Transp Res Procedia 16:378–393
6. Durand B, Gonzalez-Feliu J (2012) Urban logistics and e-grocery: have proximity delivery services a positive impact on shopping trips? Procedia-Social and Behavioral Sciences 39:510–520
7. Wang XC, Zhou Y (2015) Deliveries to residential units: a rising form of freight transportation in the US. Transp Res Part C: Emerg Technol 58:46–55
8. Agatz N, Campbell AM, Fleischmann M, Savels M (2008) Challenges and opportunities in attended home delivery. In: Golden B, Raghavan S, Wasil E (eds) The vehicle routing problem: latest advances and new challenges. Operations Research/Computer Science Interfaces, vol 43. Springer, Boston, MA. https://doi.org/10.1007/978-0-387-77778-8_17
9. Bjerkan KY, Bjørgen A, Hjelkrem OA (2020) E-commerce and prevalence of last mile practices. Transp Res Procedia 46:293–300
10. Allen J, Piecyk M, Piotrowska M, McLeod F, Cherrett T, Ghali K, …, Wise S (2018) Understanding the impact of e-commerce on last-mile light goods vehicle activity in urban areas: the case of London. Transp Res Part D: Transp Environ 61:325–338
11. Visser J, Nemoto T, Browne M (2014) Home delivery and the impacts on urban freight transport: A review. Procedia-Soc Behav Sci 125:15–27
12. Bookbinder JH, Tan CS (2003) Comparison of Asian and European logistics systems. Int J Phys Distrib Logist Manag 33(1):36–58
13. Suel E, Polak JW (2017) Development of joint models for channel, store, and travel mode choice: Grocery shopping in London. Transp Res Part A: Policy Pract 99:147–162
14. Tonn BE, Hemrick A (2004) Impacts of the use of e-mail and the Internet on personal trip-making behavior. Soc Sci Comput Rev 22(2):270–280
15. Weltevreden, J. W., & RIETBERGEN, T. V.: E-shopping versus city centre shopping: The role of perceived city centre attractiveness. Tijdschrift voor economische en sociale geografie, 98(1), 68–85 (2007)
16. Rotem-Mindali O, Weltevreden JW (2013) Transport effects of e-commerce: what can be learned after years of research? Transportation 40(5):867–885
17. Cao X, Chen Q, Choo S (2013) Geographical distribution of e-shopping: An application of structural equations models in the twin cities. In: 92nd Transportation Research Board annual meeting, Jan 2013 Washington DC, USA. Transportation Research Board of the National Academies
18. Comi A, Nuzzolo A (2016) Exploring the relationships between e-shopping attitudes and urban freight transport. Transp Res Procedia. Elsevier 12:399–412. https://doi.org/10.1016/j.trpro.2016.02.075
19. Comi A (2020) A modelling framework to forecast urban goods flows. Res Transp Econ 80:100827. https://doi.org/10.1016/j.retrec.2020.100827
20. Crocco F, Eboli L, Mazzulla G (2013) Individual attitudes and shopping mode characteristics affecting the use of e-shopping and related travel. Transp Telecommun J 14(1):45–56
21. Lee RJ, Sener IN, Handy SL (2015) Picture of online shoppers: specific focus on Davis, California. Transp Res Record 2496(1):55–63
22. Swinyard WR, Smith SM (2003) Why people (don't) shop online: a lifestyle study of the internet consumer. Psychol Market 20(7):567–597
23. Zhou Y, Wang XC (2014) Explore the relationship between online shopping and shopping trips: An analysis with the 2009 NHTS data. Transp Res Part A: Policy Pract 70:1–9
24. Farag S, Schwanen T, Dijst M, Faber J (2007) Shopping online and/or in-store? A structural equation model of the relationships between e-shopping and in-store shopping. Transp Res Part A-Policy Pract 41:125–141

25. Gonzalez-Feliu J, Peris-Pla C (2018) Impacts of retailing land use on both retailing deliveries and shopping trips: modelling framework and decision support system. IFAC-PapersOnLine 51(11):606–611
26. Hsiao MH (2009) Shopping mode choice: physical store shopping versus e-shopping. Transp Res Part E: Logist Transp Rev 45(1):86–95
27. Zhai Q, Cao X, Mokhtarian PL, Zhen F (2016) The interactions between e-shopping and store shopping in the shopping process for search goods and experience goods. Transportation. https://doi.org/10.1007/s11116-016-9683-9
28. Mokhtarian PL, Tang WL (2013) Trivariate probit models of pre-purchase/purchase shopping channel choice: Clothing purchases in northern California. In Choice modelling. Edward Elgar Publishing, New York
29. Nuzzolo A, Comi A, Ibeas A, Moura JL (2016) Urban freight transport and city logistics policies: Indications from Rome, Barcelona, and Santander. Int J Sustain Transp 10(6):552–566
30. Rohm AJ, Swaminathan V (2004) A typology of online shoppers based on shopping motivations. J Bus Res 57(7):748–757
31. Smith N, Ferreira L, Marquez L (2002) E-business impacts for urban freight: results from an Australian study. In: 81st Annual Meeting, US Transportation Research Board, 2002-01-01
32. Gevaers R, Van de Voorde E, Vanelslander T (2011) Characteristics and typology of last-mile logistics from an innovation perspective in an urban context. City distribution and urban freight transport: multiple perspectives. Edward Elgar Publishing, pp 56–71
33. Morganti E, Dablanc L, Fortin F (2014) Final deliveries for online shopping: The deployment of pickup point networks in urban and suburban areas. Res Transp Bus Manag 11:23–31
34. Morganti E, Seidel S, Blanquart C, Dablanc L, Lenz B (2014) The impact of e-commerce on final deliveries: alternative parcel delivery services in France and Germany. Transp Res Procedia 4:178–190
35. Naumov V, Kholeva O (2017) Forming the strategies of sustainable development of freight forwarders at transportation market. Naukovyi Visnyk Natsionalnoho Hirnychoho Universytetu 3:129–134
36. Macioszek E, Sierpiński G, Staniek M (2017) Analysis of trends in development of freight transport logistics using the example of Silesian Province (Poland)—a case study. Transp Res Procedia 27:388–395
37. Tubis A, Werbińska-Wojciechowska S (2018) The scope of the collected data for a holistic risk assessment performance in the road freight transport companies. In: Zamojski W, Mazurkiewicz J, Sugier J, Walkowiak T, Kacprzyk J (eds) Advances in dependability engineering of complex systems. Advances in intelligent systems and computing, vol 582. Springer, Cham, pp 450–463
38. Sierpiński G, Macioszek E (2020) Equalising the levels of electromobility implementation in Cities. In: Mikulski J (ed) Research and the future of telematics, communications in computer and information science, vol 1289. Springer, Heidelberg, pp 165–176
39. Iwan S, Kijewska K, Lemke J (2016) Analysis of parcel lockers' efficiency as the last mile delivery solution–the results of the research in Poland. Transp Res Procedia 12:644–655

The Principles and Methods of Locating Logistics Centers in Transport Networks

Elżbieta Macioszek

1 Introduction

Warehouse logistics is a key solution for the rapidly growing economy of each country, which is why its functioning and solutions are often the subjects of multifaceted research aimed at improving the increasing supply for specific products, services and the increasingly diversified needs of recipients. When analyzing the processes taking place in this field, every effort is made to facilitate and reduce the costs of distributing products in the supply chain to the customer without lowering their quality, which is very important in the context of competitiveness. The possibilities of further dynamic development of this area of the economy are very large because the necessary tools supporting faster and cheaper operation are being developed at the same time, thanks to which storage, transport and all other related aspects may become more efficient and profitable.

The Logistics Center (LC) as defined by the European Association of Logistics Centers Europlatforms [1] is located in a defined area where all activities are related to transport, logistics and distribution of both domestic and international goods. This applies to domestic and international transit. These activities are performed by operators on a commercial basis, who may be owners or tenants of buildings and facilities (warehouses, distribution centres, offices, transport services, etc.). The LC should also be equipped to carry out these operations and, if possible, should include public services for staff and equipment for users. In order to encourage intermodal transport when transhipping goods, a logistics center should be served by different modes of transport (road, rail, sea, inland waterway, air). To ensure synergy and commercial cooperation, it is important that LC is managed in a single and

E. Macioszek (✉)
Faculty of Transport and Aviation Engineering, Department of Transport Systems, Traffic Engineering and Logistics, Silesian University of Technology, Katowice, Poland
e-mail: elzbieta.macioszek@polsl.pl

G. Sierpiński and E. Macioszek (eds.), *Decision Support Methods in Modern Transportation Systems and Networks*, Lecture Notes in Networks and Systems 208,
https://doi.org/10.1007/978-3-030-71771-1_10

neutral legal body (preferably by a public-private partnership). Finally, LC must meet European standards to ensure quality within the framework required for commercial and sustainable transport and warehousing chain solutions.

Logistics centres are considered good investments stimulating the local economy and activating new ventures related to the influx of new economic entities. As a result of global economic processes, the emergence of distribution and logistics, centers have become a common phenomenon, and the most popular types of functional and spatial concentration are distribution centers and logistics parks consisting of many such structures located in one place, where the entire accompanying infrastructure is created around them. The primary goal of logistics management is to reduce the cost of end products and services and to increase the competitiveness of enterprises.

In practice, there are many classifications of logistics centers. First of all, LC can be classified depending on the scope of its functions into LC, which offers many services, including handling various loads, logistics parks grouping producers and distributors, distribution centers, warehouse centers, transport terminals providing only transport services. Another classification takes into account the scale and scope of the impact of LC and, according to it, LC is divided into LC of international, regional, local importance, universal support for various types of goods, and LC dedicated to specialized groups of goods. On the other hand, due to their location in the LC space, they are divided into LC engaged in a specific territory and having a specific logistics infrastructure, virtual, distributed, operating on the internet, in cyberspace on an electronic platform, and gathering suppliers and companies offering supplementary services in a selected region. In modern, comprehensive LC, the service recipient has access to at least two modes of transport and load handling techniques are used in conventional and combined transport.

In recent years, there has been a significant increase in trade exchange in the EU countries, and thus an increase in the mass of transported goods and transport efficiency. Road transport has a significant share in the transport work performed in the transport of goods, which for years has been perceived as having a negative impact on the natural environment, overloading the road transport network and generating the largest external costs of transport [2–7]. In this situation, one of the priorities of the EU transport policy is the development of intermodal transport and the transfer of as much cargo as possible to more environmentally friendly means of transport and waterways. Intermodal transport should allow for timely inclusion, fast and safe delivery of cargo to its destination, regardless of which cargo unit is carried and which mode of transport.

The need to build logistics centers and parks in Poland, including those of international importance, results from EU regulations. These regulations provide for an increase in the accessibility of the Polish transport market and the modernization of the transport infrastructure, taking into account EU standards. International cooperation in the development of the accessibility of logistics services is based primarily on the TEN-T network (Trans European Transport Networks), which is currently the basic structure of the European transport network.

In Poland, following the footsteps of other countries, there is a visible tendency for large logistics operators to locate logistic centers and parks on the outskirts of the cities, which allows the construction of not only larger facilities of this type as well as clusters of various types of activity, but also allows for quick connection to better quality and denser network of roads and highways in Poland. Such activities, such as the location of distribution centers and logistic parks in strategic locations from the point of view of transport, near highways, airports and seaports, provide great opportunities for the distribution of goods. The development of this area of economy in Poland in recent years has been very visible, as evidenced by the increase in warehouse space from approximately 5 million m^2 in 2008 to over 16 million m^2 in 2018, however, at the same time, the decline in construction and rental of small-area warehouses is visible. In Europe, Poland has become one of the leading countries in terms of the construction and rental of warehouse space, as well as in terms of share in road transport. Over the last dozen or so years, rapid development of LC has been visible in the western part of Poland than in the eastern part. However, along with the development of the road and motorway network, this situation is slowly changing and developers are making more and more investments also in this area of Poland. Each of these investments is more and more technologically developed in many respects, starting with environmental protection and ending with full automation in warehouses. Automation is also starting to play an increasingly important role in the case of vehicles [8, 9]. The paper presents the issues of logistics centers in terms of legal requirements for the functioning of logistics centers, rules and methods of locating such facilities in the area of transport network, which were supplemented by an analysis of the functioning of logistics centers in Poland in 2008–2018.

2 Legal Requirements for Logistics Centers

The functioning of LC is directly and indirectly regulated by a number of applicable legal acts, regulations and conventions. They can be divided into:

- Regulations of the European Parliament and the Council of Europe,
- International conventions,
- International agreements,
- Polish laws and regulations.

Among the regulations of the European Parliament are the following most important regulations:

- Regulation No. 1071/2009 of the European Parliament and of the European Council establishing common rules concerning the conditions for the exercise of the occupation of road transport operator. It regulates access to the profession of road transport operator and the pursuit of this profession and applies to all entrepreneurs established in the community,
- Regulation No. 1072/2009 of the European Parliament and of the European Council of 21 October 2009 on common rules for access to the international road haulage market. It applies to the international road haulage for hire and

reward on journeys within the territory of the community (in the case of carriage from a member state to a third country and vice versa, it applies to the part of the journey on the territory of a member state through which the transit takes place),

- Regulation No. 561/2006 of the European Parliament and of the European Council of 15 March 2006 on the harmonization of certain social legislation relating to road transport. This is because it lays down rules on driving time, breaks and rest periods for drivers engaged in road transport,
- European Council Regulation No. 3821/85 of 20 December 1985 on recording equipment in road transport. This regulation specifies uniform requirements for recording devices, their construction, installation, use and control rules,
- Regulation 1073/2009 of the European Parliament and of the European Council No. 1073/2009 of 21 October 2009 on common rules for access to the international market of coach and bus services and amending Regulation No. 561/2006 (Journal of Laws No. 300, p. 88, as amended),
- European Council Regulation No. 684/92 of 16 March 1992 on common rules for the international carriage of passengers by coach and bus (Official Journal of the EU No. 74, p. 1 as amended),
- European Council Regulation No. 881/92 of 26/03/1992 on the access to the road haulage market in the community, into or from the territory of a member state or in transit through one or more member states (Official Journal of the EU No. 95, p. 1 as amended).

In turn, the most important international conventions in this area include:

- The Convention on the Contract for the International Carriage of Goods by Road signed in Geneva on 19 May 1956. It regulates the terms of contracts for the international carriage of goods by road, in particular the documents used for this carriage and the carrier's liability. Poland ratified it in 1962,
- The Customs Convention on the International Transport of Goods under Cover of TIR Carnets of 14 November 1975, which allows the simplification of procedures at border crossing points and at customs offices, in order to increase the efficiency of road transport,
- The Convention on a Common Transit Procedure, which facilitates border-crossing transport operations. It is used when at least one border is crossed between the countries that have signed the convention. Poland acceded to the Convention on 30 June 1996.

The group of international agreements include:

- European agreement concerning the work of crews of vehicles engaged in international road transport, prepared in Geneva on 1 July 1970,
- The European Agreement on the International Carriage of Dangerous Goods by Road of 1957, which defines the rules of conducting business in the field of domestic and international road, rail and inland waterway transport of dangerous goods,

- Agreement on the international transport of perishable foodstuffs and on special means of transport intended for such transport (Journal of Laws of 1984, No. 49, item 254).

Among Polish laws and regulations, the following can be mentioned:

- The Act of 6 September 2001 on road transport, which specifies, inter alia, the rules for taking up and performing national and international road transport, the conditions and procedure for obtaining certificates of professional competence, as well as a list of infringements and the amount of fines for the driver, the person managing the enterprise, the person managing the transport in the enterprise and the entity performing road transport,
- Act of 20 June 1997, Road Traffic Law, which regulates, inter alia, rules of traffic on public roads, rules and conditions for admitting vehicles to the road traffic, rules and conditions of road traffic control and defines penalties for breaking the above regulations,
- The Act of 16 April 2004 on drivers' working time, which defines the working time of drivers, rules for the application of standards for driving periods, mandatory breaks and guaranteed rest periods, and the obligations of employers in the field of road transport,
- The Act of 15 November 1984, Transport Law, which specifies the transport of people and goods, performed for a fee on the basis of a contract by authorized carriers,
- The Act on drivers' working time of 16 April 2004,
- The Act of 1 July 2011 amending the Act on Road Transport and the Act on Public Roads (Journal of Laws No. 159, item 945),
- The Act of 16 September 2011 amending the Act on Road Transport and certain other acts (Journal of Laws No. 244, item 1454),
- Act of 6 September 2001 on road transport (Journal of Laws of 2012, item 1265, as amended).

3 The Principles and Methods for Selection the Location of Logistics Centers

The methods of LC location in the transport network mentioned in the scientific literature are classified according to the many criteria. One of tcriteria assumes choosing the location of LC based on three concepts. Each of them has a number of parameters decisive or influencing the final decision regarding the location of the warehouse facility or the entire LC. These are concepts for the location of LC near the Trans-European Transport Corridors running through Poland, in the vicinity of urban areas and in seaports. Factors influencing the choice of LC location can generally be divided into factors [10, 11]:

- natural—raw materials, plots of land, ecological barriers,

- economic and technical—sales market, workforce, agglomeration benefits,
- socio-political—social, political, strategic and military.

In turn, taking into account the criterion of the measurability of the location factors, the following factors affecting the location of LC can be distinguished [12, 13]:

- immeasurable factors—the quality of the anthropogenic and natural environment, participation of citizens in public activities, local patriotism, etc.,
- measurable factors—the geographical location of the region and its topographic features, access to the broadly understood infrastructure, e.g. educational, scientific, legal, administrative, as well as the efficiency and quality of the functioning of this infrastructure, the location and connection with an external network of technical infrastructure services, the size, quality and diversity of capital structure, size and type of ownership of existing business entities.

In the scientific literature, can be find proposals for methods of LCs locating in a regular hierarchical transport network and most often they consist in the formulation and solution of an optimization problem [14–18]. The main problem is usually the determination of the optimal number of LCs and the assignment of customers to them in such a way that the costs of servicing the area are minimal and at the same time the requirements of the production company are met, and on the other hand, the technical and economic possibilities of logistics service providers.

When planning the location of the new LC, preference is given to undeveloped areas very well connected with the existing transport network of various modes of transport with simultaneous access to media with the possibility of easy investment, as well as the possibility of further development in the adjacent areas. With the commercial nature of CL's operations, the choice of a location is mainly determined by such factors as:

- human capital potential,
- purchasing power on the part of the population,
- the level of accessibility of the workforce,
- the level of attractiveness,
- the level of communication infrastructure,
- the level of land prices,
- the level of value of the logistics market.

The factors influencing on the choice of location may also be innovation, the structure of the entrepreneur's activity or different type of infrastructure around the area in which LC will be built. The attractiveness of the region is measured on the basis of the possibilities of communication, development of infrastructure, especially the proximity of expressways and highways. These factors also affect the performance of a given LC.

Often, logistic centers and parks are located near airports or near railway terminals or port terminals. The human resources of the regions also have a large impact on the possibilities of choice, although not all of them are fully used, as can be seen in the example of eastern Poland, where such resources exist, but in these regions is less industrialized, the network of roads and highways is less developed and, importantly, a large part of society lives in small towns and villages and they are permanently associated with agriculture, which effectively reduces the amount of free human resources needed to work in LCs.

In turn, according to [19–23] the main factors in selecting the location of a LC are economic criteria, such as the value of the logistics market, attractiveness of the investment project for investors, land prices and accessibility, accessibility of transport and communication infrastructure, availability of labour and the level of education of employees, the purchasing power of the population, etc. On the other hand, factors such as environmental damage caused by road transport, unused transport potential of rail transport, road congestion, etc., are of little importance to the investors, provided that there is no state intervention through legal or financial regulations that actually change the cost relationships between the various modes of transport.

The absorptive power of the markets also contributes to the choice of LC location and, as can be seen, for example, in the western region of Poland, such spaces are more likely to be purchased or rented than in the eastern part of Poland. However, the East of Poland, has great potential and all predictions point to this direction as the location where more and more logistics investments will be made. Not only the demographic factor plays a role, but also the fact that each newly built LC is more modern than the previous one, which means that other factors taken into account by potential buyers of warehouse space are, for example, savings in operating costs through lower electricity consumption, heating and modern logistics solutions such as like warehouse automation, which is of great importance nowadays due to the demographic crisis.

When analyzing the location of LC as an enterprise, the factors determining the location should include [10]: transport, raw material base, sales market, infrastructure development (possibility of joint use of infrastructure by a group of economic entities), labor resources (possibilities of obtaining employees), energy base, water resources and agglomeration.

4 The Analysis of the Operation of Logistics Centers in Poland in 2008–2018

The operation of logistics centers in Poland in 2008-2018 was analyzed. The supply in the warehouse market in the years preceding 2008 was growing. Since 2008, due to the global crisis, there has been a decrease in supply. An upward trend in supply was recorded again from 2010, which continued until 2014. It should be noted that during the crisis, developer companies had limited access to finance. Poland's accession to the EU in 2004 resulted in a dynamic increase in interest in warehouse space, which translated into the creation of numerous logistics centers and parks. In Poland, in

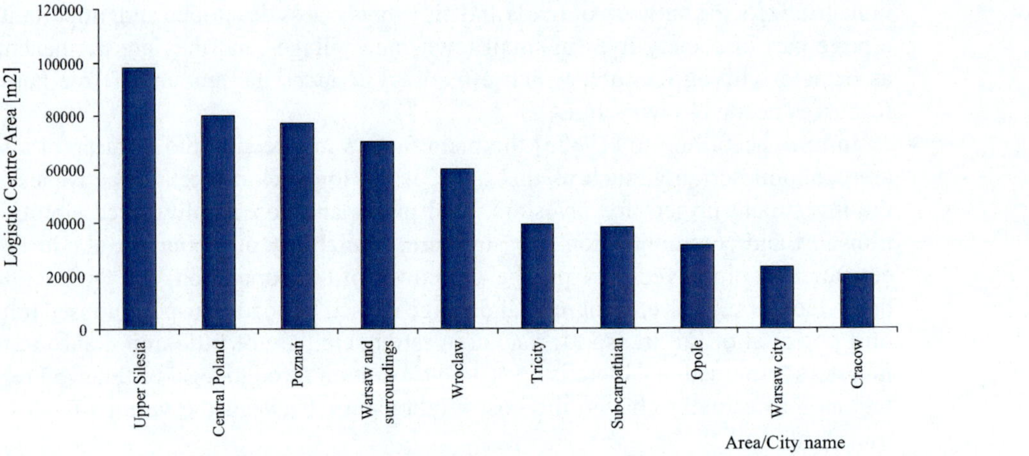

Fig. 1 The opened logistic centers area in Poland in 2018. *Source* Own researdch based on [24]

2018, over 500,000 m^2 of newly built warehouse space was commissioned (Fig. 1). The smallest amount of warehouse space was built in Cracow, Warsaw and Opole.

In turn, on the basis of Fig. 2a, it can be concluded that the region of Warsaw and the Warsaw surrounding area is the most developed area in warehouses and LCs.

Also, when analyzing the number of individual LCs in individual voivodships of Poland (Fig. 2b), it can be concluded that the largest number is in the Masovian voivodship. In research studies and many statistics, this region is often additionally divided into two subregions: Waraw City and Warsaw Surroundings. On the Fig. 3, the increase in warehouse space in Poland in 2008-2018 is presented. The data presented on the Fig. 3 show that the warehouse space in Poland has increased significantly in recent years.

Due to the strategic location and well-developed road infrastructure, as well as the presence of transhipment terminals in the vicinity, the areas of central Poland and Upper Silesia enjoy the great interest in the logistics sector. On the example of several large regions of Poland, the development, demand and supply as well as rental prices

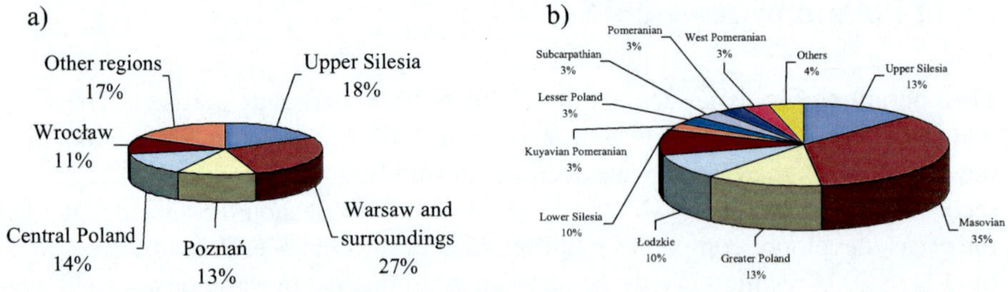

Fig. 2 The share of individual regions of Poland in the warehouse space market in 2018. **a** by region, **b** by voivodeships. *Source* Own researdch based on [25, 26]

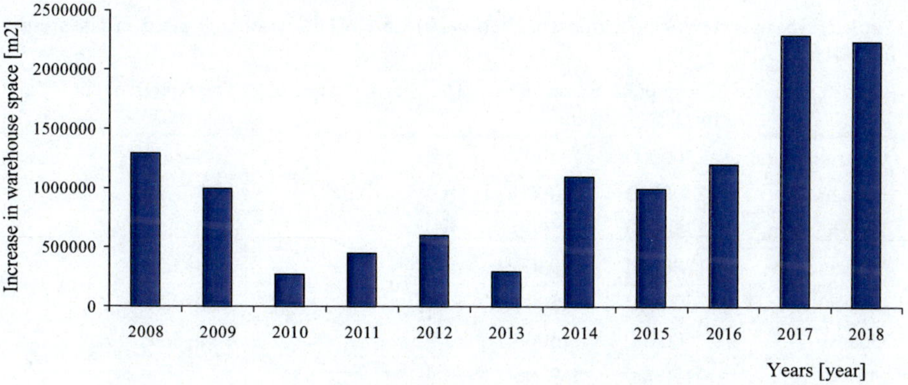

Fig. 3 The increase in warehouse space in Poland in 2008-2018. *Source* Own research based on [27]

Table 1 The overview of the regional markets in 2008. *Source* Own research based on data presented in [28]

City/Region	The supply [m^2]	Free area [m^2]	The vacancy rate [%]	The effective rent [€/m^2]
Central Poland	779500	92200	11.8	3.2–3.6
Poznań	730000	102300	14.0	3.2–3.6
Upper Silesia	867460	54400	6.3	3.2–3.7
Wrocław	542000	125300	23.1	3.3–3.7
Cracow	35900	10500	29.2	4.2–4.5
Gdańsk	82000	0	0.0	3.5–3.9
Sum	3036860	384700	22.2	–

for one square meter of warehouse space were analyzed. Based on the data contained in Tables 1 and 2, it can be concluded that the number of used warehouse space has increased several times over the past ten years, with a simultaneous decrease in the vacancy rate. The construction of numerous new centers and logistics parks in recent years has resulted in a reduction in rental prices. Nevertheless, Cracow and its vicinity remain the most expensive region in terms of rental prices for warehouse space. It can be seen that the eastern part of Poland is practically not taken into account in most statistics and reports, due to the low density of such objects in these regions of Poland. However, attention should be paid to the dynamic development of the Podkarpackie Logistics Park in Jasionka near Rzeszów, the location of which is very rational, because in a short distance from the Podkarpackie Logistics Park in Jasionka there is the A4 motorway connecting the eastern and western borders of Poland. In addition, the construction plans include the ViaCarpatia transport corridor connecting, in the original concept, the Baltic countries with the countries of southern Europe. It is to run through countries such as Lithuania, Poland, Slovakia, Hungary, Romania,

Table 2 The overview of the regional markets in 2018. *Source* Own research based on data presented in [28]

City/Region	The supply [m^2]	Free area [m^2]	The vacancy rate [%]	The effective rent [€/m^2]
Central Poland	2700000	135000	5.0	2.40–3.60
Poznań	2000000	180000	9.0	2.75–3.40
Upper Silesia	2750000	165000	6.0	2.90–3.50
Wrocław	1700000	49300	2.9	2.90–3.50
Cracow	500000	25000	5.0	3.50–4.50
Gdańsk	500000	15000	3.0	2.90–3.50
Sum	10150000	569300	30.9	

Bulgaria and Greece. The potential of the via Carpatia route can be used in many transport relations, both inside and outside the EU, also in relation to the Balkans. In the vicinity of this park, there is also a dynamically developing airport in Jasionka. The Podkarpackie Logistics Park in Jasionka is an example of a well-thought-out location of a logistic park, which may continue to develop in the future due to a large amount of land free of buildings, as well as the possibility of cooperation with the airport and the expansion of the logistics zone with air cargo, which at present is not yet a standard in Polish conditions.

The supply on the warehouse space market is constantly growing, yet there is a shortage of warehouse space during the holiday season (f.ex. like Christmas time).

This phenomenon is caused by the fact that many companies, especially commercial ones, create the so-called buffer warehouses by storing and at the same time securing against financial losses caused by the lack of sale of selected products. In 2008-2018, the supply of warehouse space in Poland increased more than threefold (Fig. 4). At the end of 2018, there were nearly 16 million m^2 of warehouse space in Poland.

The results of cyclical research conducted in territorial units in Poland, including both small-format and large-format warehouses, show a decrease in the number and area of small-format warehouses in relation to the number of square meters of large-format spaces (Fig. 5). This is confirmed by the fact that logistics is transferred to the large logistics operators with significant financial resources. Large logistics operators most often displace small entrepreneurs or, at best, offer them a cheaper rental offer, reduce storage and internal logistics costs, which allows large and small companies to operate simultaneously at lower costs and at the same time fills the warehouse space and becomes profitable for both parties.

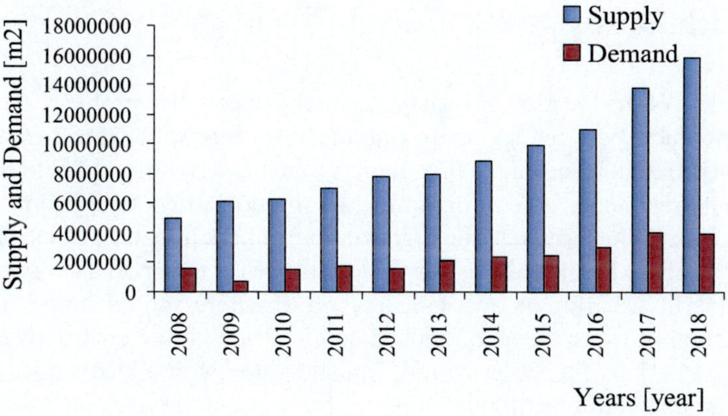

Fig. 4 Podaż i popyt powierzchni magazynowej w Polsce w latach 2008-2018. *Source* Own research based on [25]

Fig. 5 The decreases in the number of small-format stores in Poland in different voivodeships in the years 2008-2018. *Source* Own research based on data presented in [29]

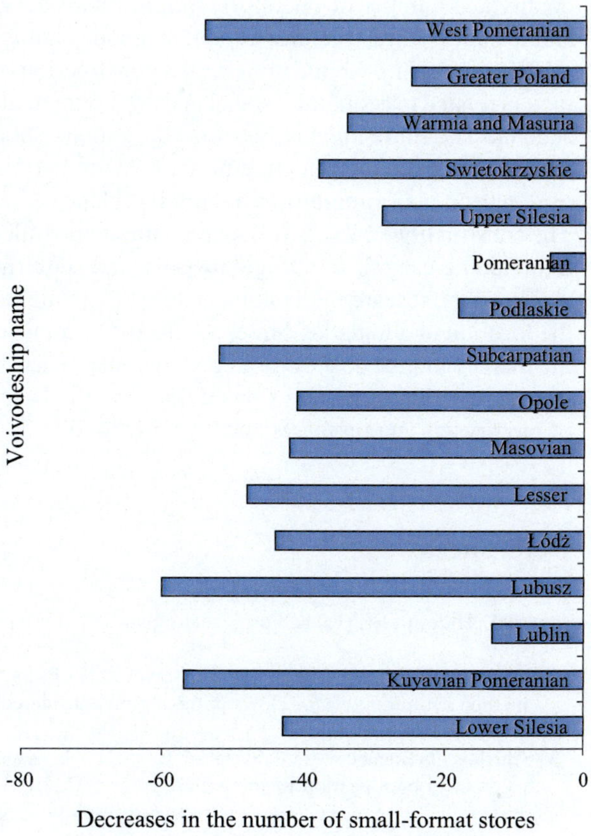

5 Conclusions

The problem of the location of logistics centers is a complex problem in terms of the flow of products between suppliers and logistics centers, between logistics centers located at different levels of the hierarchical logistics network, and between logistics centers and recipients. According to the assumptions, LC should carry out a wide range of tasks, which include the organization and coordination of loads, including on an international scale, connecting various modes of transport and organizing their interoperability, sharing its own logistics infrastructure and creating conditions for the operation of logistics operators and companies with the environment business, creating conditions for the growth of logistic value with added value, creating an area of fair competition and market prices for logistics services on a local, regional and international scale. In addition, in its activities, LC controls transport processes through their appropriate organization and coordination influencing the development of the transport industry and types of transport in the region. LC affects the shape, quality and number of operators, freight forwarders, cargo handlers, transport, line and technical infrastructure, as well as goods, storage and technical facilities.

When selecting the location for the construction of a logistics center, a number of factors related to economic, social and environmental conditions should be taken into account. The entire logistic system of the country should be taken into account, which should guarantee the effective flow of raw materials, goods, and finished products of domestic and foreign origin from the place of acquisition to final consumption. The construction of LC is still faced with many difficulties, which are mainly related to obtaining capital, a strategic investor and acquiring land. However, based on the results of the research, it is indicated that the main factors determining the choice of the location of a logistics center are access to transport infrastructure, distance from the market and the cost of renting warehouse space. The choice of location and then the construction of LC are slowed down by the lack of a sufficiently dense network of motorways, expressways and city ring roads.

References

1. Europlatdorms EEIG. The European Logistics Platforms Association http://www.europlatforms.eu/
2. Chislov ON, Bogachev VA, Zadorozhniy VM, Bogachev TV (2018) Economic-geographical method delimiting wagon flows in the region considered: model and algorithm. Transp Probl 13(2):39–48
3. Chislov O, Bogachev T, Kravets A, Bogachev V, Zadorozhniy V, Egorova I (2020) Time parameters optimization of the export grain traffic in the port railway transport technology system. In: Sierpiński G (ed) Smart and green solutions for transport systems. AISC, vol 1091. Springer, Cham, pp 126–137
4. Danchuk V, Bakulich O, Svatko V (2019) Identifying optimal location and necessary quantity of warehouses in logistic system using a radiation therapy method. Transport 34(2):175–186

5. Macioszek E (2018) First and last mile delivery—problems and issues. In: Sierpiński G (ed) Advanced solutions of transport systems for growing mobility, vol 631. AISC. Springer International Publishing, Switzerland, pp 147–154
6. Macioszek E (2019) Freight transport planners as information elements in the last mile logistics. In: Sierpiński G (ed) Integration as solution for advanced smart urban transport systems. AISC, vol 844. Springer International Publishing Switzerland, pp 242–251
7. Macioszek E (2019) Changes in values of traffic volume—case study based on general traffic measurements in Opolskie Voivodeship (Poland). In: Macioszek E, Sierpiński G (eds) Directions of development of transport networks and traffic engineering, vol 51. LNNS. Springer International Publishing, Switzerland, pp 66–76
8. Macioszek E (2019) E-mobility infrastructure in the Górnośląsko—Zagłębiowska Metropolis, Poland, and potential for development. In: Proceedings of the 5th World congress on new technologies (NewTech'19), Paper No. ICERT 108, pp. ICERT 108-1-ICERT 108-4. https://doi.org/10.11159/icert19.108. Lisbon
9. Macioszek E, Sierpiński G, Staniek M (2017) Analysis of trends in development of freight transport logistics using the example of Silesian Province (Poland)—a case study. Transp Res Procedia 27:388–395
10. Godlewska H (2001) Location of the business. Selected Issues. University of Trade and International Finance, Warsaw
11. Iskender P, Birdogan B, Mehmet T, Olker M (2016) Logistics center site selection by ANP/BOCR analysis: a case study of Turkey. J Intell Fuzzy Syst 30(4):2383–2396
12. Jędra I, Borowiak J (2010) Location determinants of logistics centers in Poland. Logistics 6:1211–1218
13. Liu X, Guo X, Zhao X (2012) Study on logistics center site selection on Jilin Province. J Softw 7(8):1799–1806
14. Jacyna M, Lewczuk K, Kłodawski M (2019) Warehouse management and designing with SIMMAG3D. Oficyna Wydawnicza Politechniki Warszawskiej, Warsaw
15. Jacyna M, Kłodawski M, Wasiak M, Żak J (2019) Planning the development of sustainable transport system with simulation tools. Oficyna Wydawnicza Politechniki Warszawskiej, Warsaw
16. Jacyna M, Krześniak M (2018) Computer support of decision-making for the planning movement of freight wagons on the rail network. In: Macioszek E, Sierpiński G (eds) Recent advances in traffic engineering for transport networks and systems, vol 21. LNNS. Springer International Publishing, Cham, pp 225–236
17. Izdebski M, Jacyna M, Klimek P, Jacyna-Gołda I, Vasek R (2019) Decision-making problems of designing database in the aspect of planning transport of cargoes by intermodal transport. In: MATEC Web of Conferences, vol 294. MATEC Web Conferences, pp 1–6
18. Tu CS, Chang CT, Chen KK, Lu HA (2010) Applying an AHP—QFD conceptual model and zero-one goal programming to requirement-based site selection for an Airport Cargo Logistics Center. Int J Inf Manag Sci 21:407–430
19. Fechner I (2004) Logistics centers. Goal—Realization—Future. Institute of Logistics and Warehousing, Poznań
20. Eryuruk SH, Kalaoglu F, Baskak M (2013) Comparison of logistics and clothing sectors for a logistics center site selection using AHP. Fibres Text Eastern Europe 2(98):13–18
21. Sładkowski A, Cieśla M (2018) Analysis and development perspective scenarios of transport corridors supporting Eurasian trade. In: Sładkowski A (ed) Transport systems and delivery of Cargo on East-West Routes, vol 155. SSDC. Springer International Publishing, Cham, pp 71–119
22. Cieśla M (2015) Outsourcing strategy selection for transportation services based on the make-or-buy decision. Transp Probl 10(2):91–98
23. Mrówczyńska B, Cieśla M, Król A, Sładkowski A (2017) Application of artificial intelligence in prediction of road freight transportation. Promet-Traffic Transp 29(4):363–37
24. Jones Lang La Salle, Magazyny.pl IV kw. 2012, https://magazyny.pl/media/plugins/report_file/Rynek_powierzchni_magazynowych_w_Polsce_2012_final_1.pdf

25. Cushman & Wakefield Polska, https://www.cushmanwakefield.com/pl-pl/poland
26. Jartom Real Etate, http://www.jartom.com/Raport-rynku-powierzchni-magazynowych-2018-2019
27. Cushman & Wakefield, https://industrial.pl/aktualnosci/raporty/194-marketbeat-rynek-magazynowy-w-polsce-i-kwartal-2019-r-raport
28. Warehouse Market, https://warehousemarket.pl/wp-content/ uploads/2019/02/Savills_Market-in-Minutes_Rynek-magazynowy-w-Polsce_07-02-2019
29. Central Statistical Office, http://stat.gov.pl/

The Comparative Method of Assessing City Logistics Measure

Mariia Olkhova, Dmytro Roslavtsev, and Andrii Galkin

1 Formulation of the Problem

Ensuring a sustainable transport system of the city is one of the important tasks of the local authorities. Issues of sustainability and mobility are new for Ukrainian cities, both for local governments and for other users of transport infrastructure. At the same time, there is a growing interest and the need to use the best world experience in city logistics activities in the largest cities of Ukraine. The issue of the sustainable transport system in Ukraine is beginning to be implemented only in the context of the organization of road traffic and public transportation, while the issue of efficient city logistics and ensuring high environmental quality standards have not yet become relevant. Dynamic cities of Ukraine, which implement a comprehensive approach to the transport system, are Kharkiv, Kyiv, Lviv, Rivne. At the same time, there is an obvious tendency to increase population density, increase retail stores, increase the number of hotel, restaurant, and catering, e-commerce operations, construction of facilities, that is all those activities that lead to increased city freight demand. Due to this trend, it is considered relevant to study the impact of city logistics activities on the direct participants in the supply chain. It is the shipper, the carrier, and the consumer who are currently almost out of control, have freedom of movement, loading, and unloading in the city infrastructure. But generally, the introduction of urban logistics measures obviously will affect the participants in the supply chain. Moreover, there is

M. Olkhova (✉) · D. Roslavtsev · A. Galkin
Transport Systems and Logistics Department, O.M. Beketov National University of Urban Economy in Kharkiv, Kharkiv, Ukraine
e-mail: olhovamv@gmail.com

D. Roslavtsev
e-mail: d.roslavtsev@gmail.com

A. Galkin
e-mail: galkin.tsl@gmail.com

G. Sierpiński and E. Macioszek (eds.), *Decision Support Methods in Modern Transportation Systems and Networks*, Lecture Notes in Networks and Systems 208,
https://doi.org/10.1007/978-3-030-71771-1_11

a high probability that the first city logistics measures will be regulatory in Ukraine. Such as time windows, truck ban in the city center, low emissions zone, etc. Therefore, this study aims to analyze the regulatory city logistics measures in terms of the impact on the transport operator. It is considered relevant to conduct a comparative analysis of the delivery process with and without road restrictions.

2 Analysis of Recent Research and Publications

A vast deal of research is devoted to the study of city logistics activities. Many scientific papers and case studies have been developed thanks to the policy of the European Union on the need to follow Sustainable Urban Mobility Plans in the framework of projects BESTUFS, CIVITAS, SUTP, NOVELOG, SULPiTER, ENCLOSE, and others [1–6]. One of the global goals of the projects is to reduce the negative impact of urban freight transport. Reduction of the negative impact can be achieved through efficient organization of freight transportation, the introduction of new technologies, implementation of Intelligent Transportation Systems in the cities, using low emission vehicles, etc. The implementation of such measures requires a clear interaction of all stakeholders (supply chain participants, city authorities, citizens, etc.), but to achieve such interaction in practice, most commonly, becomes a very sophisticated task. Since there is a conflict between stakeholders [7, 8].

The analysis of a large number of city logistics projects based on the taxonomy approach is presented in the paper [9]. The differences and similarities of the projects are described. A detailed analysis of possible approaches to the classification of city logistics measures and their classification proposed are presented by prof. Russo and prof. Comi [10]. Another offered classification of city logistics measures consists of four types such as material infrastructure, immaterial infrastructure, equipment, and governance measures. The relationships among strategies, measures, and planning levels of city logistics measures are summarized by prof. Comi and prof. Nuzzolo [11]. They offered to highlight measures depending on the type of flow—restocking and shopping. Taking into account that all possible city logistics measures must be compatible with sustainable logistics principles, authors distinguished urban logistics measures by 3 pillars of sustainability—economic (traffic congestion, trip length, delivery time, infrastructures cost), environmental (reduction of pollutants, reduction of noise, habitat loss) and social (reduction of interferences among segment of urban mobility, reduction of operating vehicles, reduction of road accidents, the liveability of the city) [12].

A crucial point of city logistics measures is assessing. The papers contain studies related to ex-post and ex-ante assessments of urban freight transport policies. But in paper [13] authors claimed that a methodology to support the measures prior to implementation is thus crucial to evaluating ex-ante the possible effects. Prof. Comi and prof. Nuzzolo made a huge contribution to city logistics planning and modeling based on ex-ante assessment. Their paper [13] notes, that in order to implement the most effective city logistics measures, the choice of a set of measures should be based on a design scenario implementation process, which consists of several steps able to: reveal the current critical issues through specific surveys, share objectives and strategies as an optimal compromise among the different actors involved, define the results indicators, find suitable measures and set up a new city logistics scenario, define models to simulate the current scenario and assess the future one, and assess ex-ante the new scenario by estimating impacts and system performance, and compare them with the set of their given pre-fixed values.

An example of separation of measures concerning ex-ante assessment of urban freight transport policies is divided into the following classes: freight traffic regulation, physical infrastructure, Intelligent Transportation Systems (ITS), loading units, and vehicles [13]. In article [14] several features of transport and the negative consequences were highlighted and correlated with measures of urban logistics to reduce the negative impact. A review on identifying the impact areas of the most implemented city logistics measures which were divided into 6 clusters, attempting to correlate the main attributes of each measure with expected core impacts to facilitate the decision-making process is conducted in the paper [15].

A Multi-stakeholders Multi-Criteria Decision Analysis tool to provide directions considering all the different components of sustainability under the SULPiTER project has been developed. Thus, it's possible to calculate Logistics Sustainability Index (LSI) due to aggregate normalized values of indicators into a unique index able to assess the city logistics measure's impacts over a given impact area, and to aggregate different indexes to assess measure's implementation [5]. The study [16] also examined the use of a multi-criteria model that would take into account the interests of each of the stakeholders, creating training models and a model for the routing and planning of a vehicle with a predicted time window that used the Q-learning method, a reinforcement learning technique when constructing a training model. Specific policy measures for improving urban logistics procedures in small and medium-sized cities through Multi-Criteria Analysis as part of the development of a Sustainable Urban Logistics Plan (SULP) for the city of Serres in Greece have been evaluated in the paper [17].

Large research is devoted to the assessment of the effect of city logistics measures connected with the regulatory and economic point of view by city authorities [18–20]. For assessing the implementation of the Limited Traffic Zone in Rome an agent-specific approach is applied [19]. This approach allowed to increase decision-makers' awareness and help taking better decisions. Logistics measures on the part of carriers are presented in the article [21]. They, like other participants in the transport process, can take an active part in solving problems in the field of urban freight transport. There are 3 main directions for carriers to improve transport mobility in the cities: vehicle innovations, i.e. engine solutions, vehicle and equipment solutions to make low-noise deliveries possible, ICT and ITS solutions that improve the efficiency of carriers' operations. The comparative analysis of using electric freight vehicles in Kharkiv (Ukraine) has been provided in the paper [22].

3 Research

3.1 Approach to Research

The chapter aims to analyze the impact of the implementation of time window and truck ban through the city center while delivering to the retail chain. The task is to minimize the impact of the implementation of city logistics measures on the carrier. The efficiency criterion is the technological performance of vehicles on the routes.

It is possible to identify several alternatives under these conditions. Firstly, it is necessary to shift the schedule of vehicles on the routes considering the service time of consumers. This shifting may cause additional costs of both time and money. Secondly, during planning routes, it is needed to avoid driving through the central streets of the city, which can be prohibited to move. The developed models of cargo transportation under such changed conditions will allow predicting the time required for the delivery within the city and the cost, due to the need to serve consumers in the new conditions of transport infrastructure.

The evaluation of the proposed models can be carried out with the help of technological vehicle performances and transportation costs. The comparative analysis is selected for the study, which allows investigation of the delivery process considering two scenarios with and without city logistics measures implementation.

The implementation of these measures leads to the need to adapt to the transport process in the city, namely the need to change the schedule of drivers and other staff. After the implementation of these measures, it is necessary to analyze positive and negative outcomes for all stakeholders.

The research was conducted in the following sequence: defining the boundaries of the system and setting constrains; design of delivery routes; study of the impact of delivery schedule changes on the delivery; study of the impact of truck ban in the city center on the delivery; evaluation of the implementation of the proposed city logistics measures.

3.2 Defining System Boundaries and Setting Constraints

Let's describe the system in detail. The object of the study is the process of transportation of food products supplied to the supermarket chain of Posad LLC in Kharkiv. Posad LLC currently has 24 supermarkets and 118 small stores. The distribution network of Posad LLC in Kharkiv is shown in Fig. 1. Suppliers of products for Posad LLC can be divided by size into large, medium, and small. The largest of them are Kharkiv Biscuit Factory, Technocom LLC, Kharkiv branch of Sun InBev Ukraine, Branch of Wimm-Bill-Dann Ukraine, Polyus LTD LLC, Kharkiv Meat Plant LLC, Olimp LLC, and others. In a further study, two suppliers were selected among them—Olimp LLC and Polyus LTD LLC.

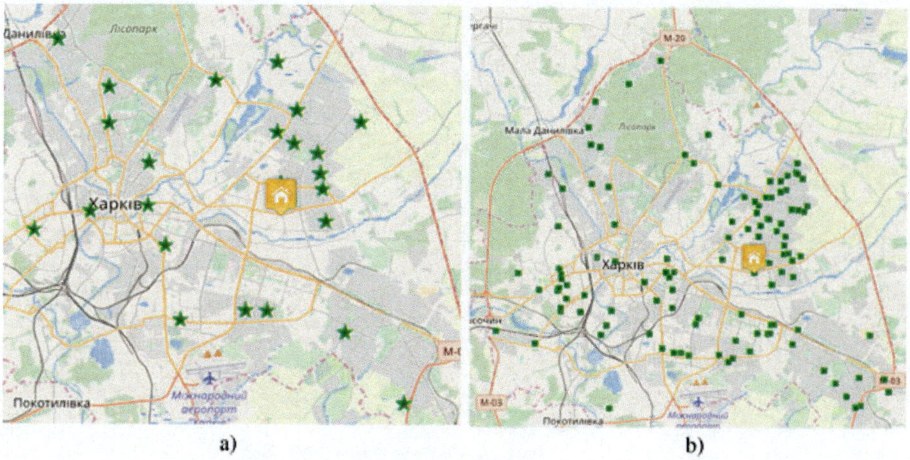

Fig. 1 The distribution network of Posad LLC in Kharkiv: **a** supermarket. **b** Small stores

Olimp LLC is a leading distributor of groceries in the Eastern region of Ukraine. Polyus LTD LLC is a confectionery factory widely known for its range of waffle products. The range includes bars, candies, gift cakes, and waffle semi-finished products. Transportation of products is carried out by specialized transport, equipped following the requirements of sanitary rules and regulations. Transportation conditions (temperature, humidity, etc.) meet the requirements of regulatory and technical documentation for each type of food. Transportation of products of Olimp LLC and Polyus LTD LLC is carried out by vehicles with a carrying capacity from 2 tons to 5 tons. The service time lasts till 6 pm.

3.3 Route Designing

Route designing was carried out using the online server Ant Logistics [23]. The server is based on the ant colony optimization algorithm. When calculating delivery routes, the following factors can be taken into account: dimensions, load capacity, vehicle type, commodity type, weight, temporary restrictions on delivery of goods, road category, direction, speed limit, etc. [23, 24]. The efficiency criterion of the transportation process is the minimum distance of transportation provided a maximum load of vehicles.

After analyzing the market for selected products by selected suppliers, the number of cargoes to the distribution network of Posad LLC in Kharkiv was formed. Two types of vehicles were selected: refrigerators with a capacity of 2 to 3 tons, the average speed is 32 km/h, and trucks with a capacity of 3-5 tons, the average speed is 29 km/h. The indicators of the delivery routes are given in Table 1.

Table 1 Delivery routes indicators

Route	Distance, km	Time, h	Stops	Weight, kg
Polyus LTD LLC/Olimp LLC				
1	61,53/68,75	7,53/7,41	18/18	1 751/2 328
2	64,28/54,34	6,93/6,81	18/17	953/2 356
3	82,02/56,44	7,81/6,85	18/17	1 167/2 217
4	69,47/52,34	7,5/6,96	19/18	1 385/2 439
5	66,80/37,59	7,06/6,95	18/21	1 155/2 604
6	90,75/73,62	7/7,06	14/16	717/1 981
7	70,54/58,77	7,6/6,90	19/19	1 317/2 118
8	78,18/51,53	7,68/6,31	18/16	1 204/1 811

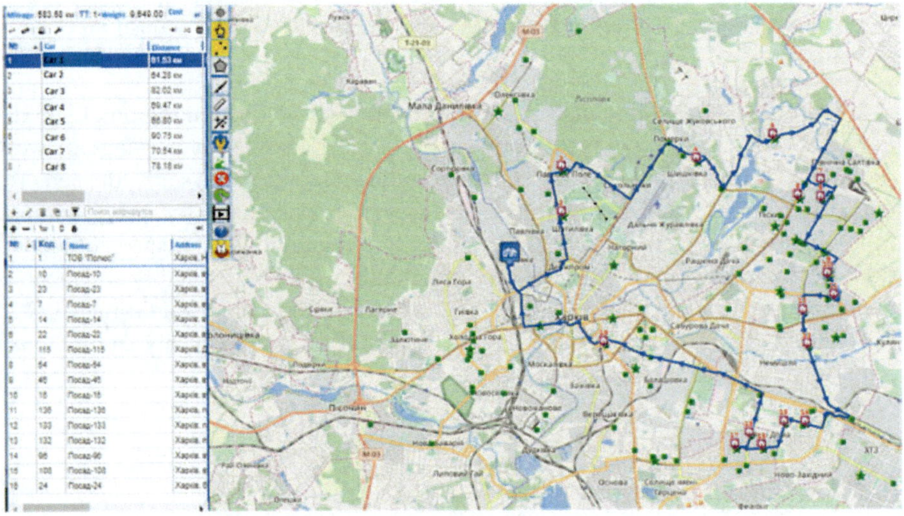

Fig. 2 Route scheme №1 during the supply of confectionery of Polyus LTD LLC

The scheme of the route №1 during the supply of confectionery of Polyus LTD LLC is shown in Fig. 2.

3.4 *Impact of Delivery Schedule Changes on the Delivery*

Due to the analysis of measures to reduce the negative impact of freight transport in European cities [1–7], it's possible to reveal measures that can be used to improve the efficiency of the transport system in Ukrainian cities. In particular, there is the problem of traffic jams. In the morning and evening, there is a maximum number of vehicles on the roads, including individual cars, trucks, and public transport vehicles.

Table 2 The total time on the routes with respect to start time

Supplier	Start time (number of routes)			
	8.00	*9.00*	*10.00*	*11.00*
Polyus LTD LLC	51,58 (8)	55,15 (8)	62,45 (9)	70,4 (13)
Olimp LLC	48,19 (8)	56,69 (8)	56,31 (8)	70,39 (10)

A possible measure to increase road capacity is to ban the movement of trucks during rush hour. As a result of this measure, the shipper, the carrier, and the consumers must adapt to these changes. How these changes will affect the performance of the transportation process needs to be established. Thus, the situation is modeled when the city authorities set restrictions on the movement of trucks in the city more than 2 tons capacity within the city during rush hours—from 07:00 to 10:00 am. It was changed the delivery time for the suppliers according to three options: the starts at 9.00, 10.00 and 11.00. Under these conditions, the total operating time of vehicles on the routes increases, Table 2.

The reason for the increase is the need to use a greater number of vehicles to deliver the same amount of cargo at the same end of working time.

3.5 *Impact of Truck Ban in the City Center on the Delivery*

Another common practice of reducing the impact of freight transport in cities is to ban trucks moving in the city center [7]. Obviously, this initiative affects the process of delivering products within the city for the carrier. Therefore, further research is devoted to this issue.

Measures are proposed to restrict the movement of trucks of more than 3 tons of capacity within the city center. Streets on which the movement of trucks is prohibited the following: Sumska Str., Pushkinska Str. (to the intersection with Vesnina Str.), Constitution Square, Rymarska Str., Maidan Svobody, Svobody Str., Pavlovskaya Square, Independence Ave., Nauki Ave. (to the intersection of Otakara Yarosha Str.), Brusatsky Descent, Korolenko Str., Trinklera Str., Darwina Str., Alchevskikh Str., Chernyshevskoho Str., Yaroslava Mudrogo Str. Let's analyze the impact of this measure in the case of different options for the ban. Four possible options are considered, during which the movement of trucks in the central part of the city may be prohibited, Fig. 3.

During the simulation, other routes for the transportation of finished products to the trade network of Posad LLC were obtained. Characteristics of these routes are described below.

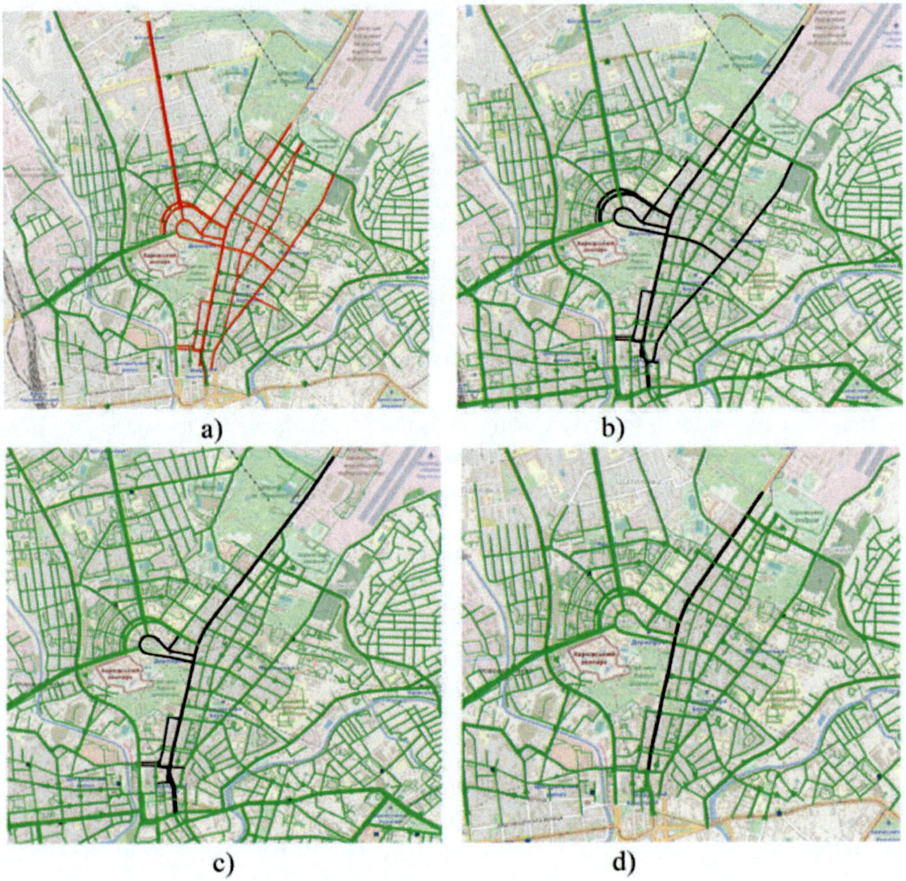

Fig. 3 Streets with limited traffic of different lengths: **a** 19.15 km, **b** 11.73 km, **c** 6.52 km, **d** 4.2 km

3.6 Evaluation of the Proposed City Logistics Measures

The considered measures in terms of technological and economic indicators have been evaluated. Economic indicators include delivery costs, which are determined based on the rate. The main factors on which the delivery tariff depends on are weight, distance, number of stops, the presence of a freight forwarder. Based on the rating, the delivery costs for each of the suppliers were calculated, depending on the amount of cargo transported, the number of vehicles, the cost and fuel consumption, mileage, Table 3.

During the introduction of restrictions in the city, the technological indicators increase, as well as increases the transportation cost, Figs. 4 and 5.

At the same time, truck ban causes a greater negative effect on the transportation process in comparison with time window increasing transportation costs. During the implementation of both measures, the trend is similar, there is a negative impact on

Table 3 The results of the calculation of the considered alternatives per year

Supplier	Indicator	Before	After schedule shifting	After truck ban
Polyus LTD LLC	Number of routes, K	2	3,25	3,25
	Mileage, K km	145,892	156,930	178,357
	Service time, day	537,2	733,3	757
	Cost, K UAH/t	174,191	194,727	196,869
Olimp LLC	Number of routes, K	2	2,5	3,25
	Mileage, K km	113,347	131,187	155,94
	Service time, day	501,9	639,1	733,22
	Cost, K UAH/t	104,1	109,265	116,903

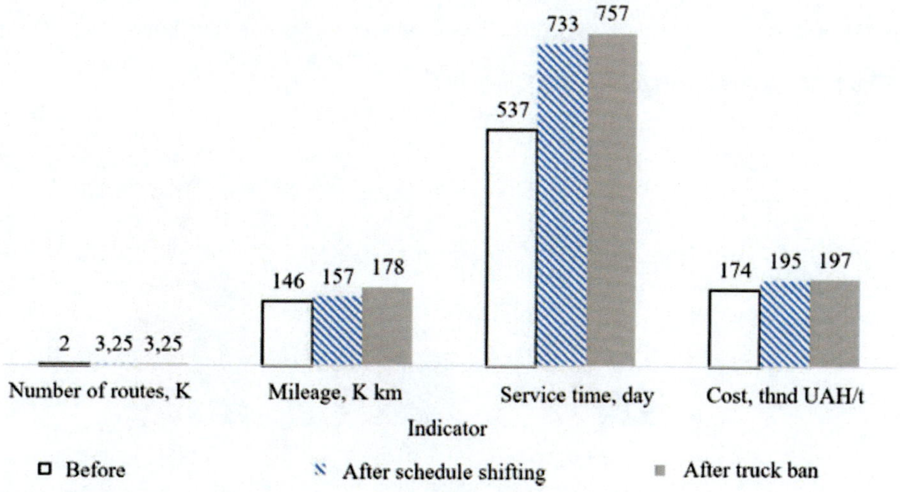

Fig. 4 Project indicators for the year, Polyus LTD LLC

the transportation process by increasing the number of routes, time, mileage, and costs. Regarding time window, the carrier has a significant increase in service time, which also leads to additional costs, Fig. 6. This is due to the need to use a greater number of vehicles to serve consumers under these conditions.

Truck ban through the city center causes an increase in mileage for vehicles. The percentage increase in mileage due to this measure is illustrated in the graph, Fig. 7.

As it can be seen there may even be a situation where truck ban can lead to a decrease in the total mileage of vehicles (if the length of the forbidden road is 4.2 km). Certainly, the factor of road congestion is also crucial to evaluate. Unfortunately, it is not taken into account in this study, but in further research it is relevant. From the preliminary results, it can be concluded that the ban on 1 km of road leads to an increase in vehicle mileage by an average of 3 km for Olympus LLC and 1.3 km for Polyus LTD LLC. These conclusions are relevant only for the companies under consideration under specific conditions of transportation.

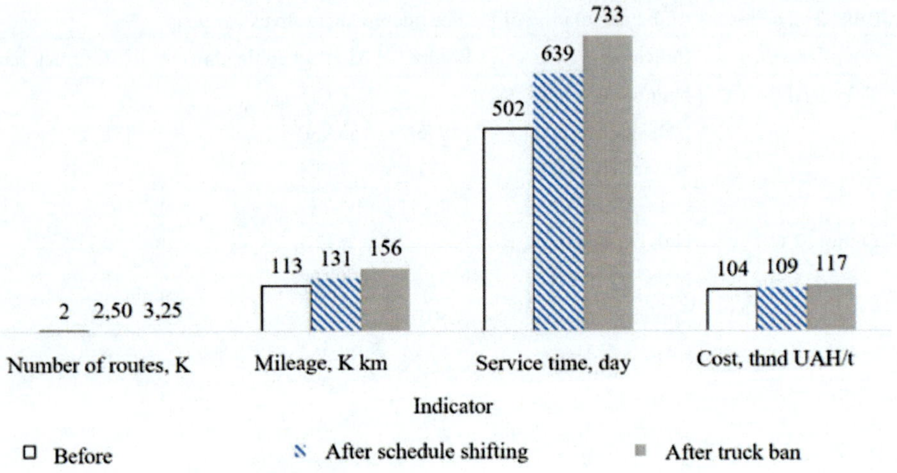

Fig. 5 Project indicators for the year, Olimp LLC

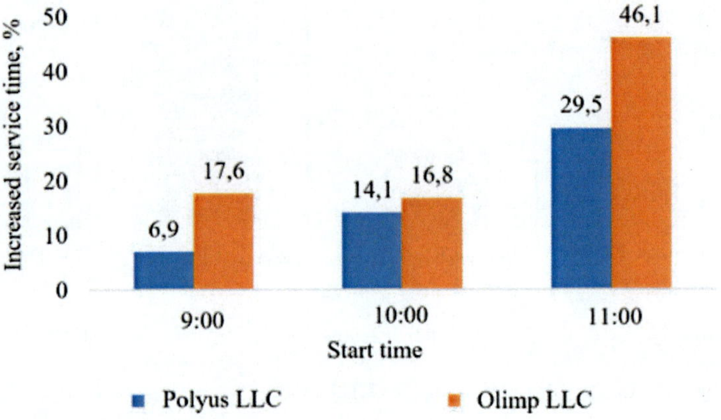

Fig. 6 Increased service time while time window

4 Conclusion

The implementation of city logistics measures in the cities of Ukraine has not yet become sufficiently relevant, both by the city authorities and by other stakeholders. Moreover, freight transport is managed through regulatory measures, such as the establishment of prohibited signs for the vehicles on the line road (more than 3.5 tons). Other city logistics measures are not currently observed. The reasons for this are, firstly, the lack of financial resources of local authorities, and secondly, the scale of the problem, i.e. there are no major problems with freight transportation compared to the cities of European Union. But there is a rapidly growing trend of increasing

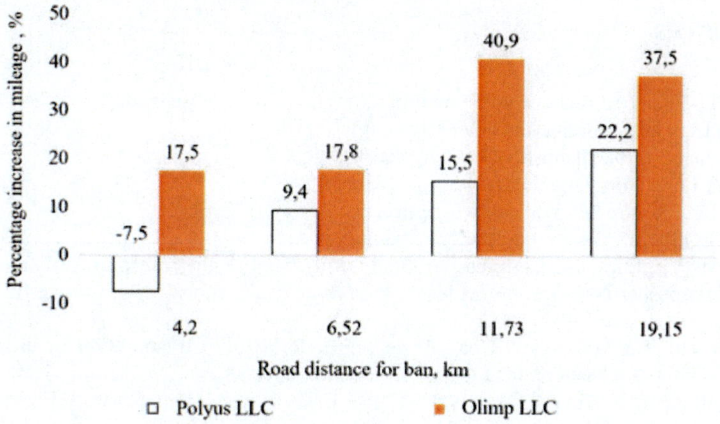

Fig. 7 Percentage increase in mileage due to truck ban

demand for freight in the city due to increasing population density, the number of small retail stores, construction activities, e-commerce services, etc.

A study of the impact of possible city logistics measures in Kharkiv on the transportation of finished products in the retail chain of grocery stores has been considered. The implementation of regulatory measures by the city authorities such as the time window and truck ban through the city center is considered and their impact on the carrier has been assessed. Delivery routes before and after implementation of city logistics measures have been designed. Thus, it is established that if the measures are implemented, there will be significant changes in the transportation process: increase the number of routes, increase the total vehicle mileage while delivering the same cargo weight, increase the required number of vehicles and operating time. All this, certainly, leads to increased delivery costs to the retail chain. But these measures can reduce the road loading in the city center, create additional space for public transport, improve the attractiveness of the city for residents and tourists.

Although the study was conducted with a lot of restrictions, the issue of freight transport regulation in the city is complex multiple tasks. Also, for the analysis of the implemented measures, it is expedient to consider the impact on the environment, changes in the city traffic, etc. On the one hand, the task of the city authorities is to ensure a sustainable transport system, on the other hand, the implementation of these regulatory measures leads to more harmful emissions into the atmosphere. Therefore, it is necessary to consider in more detail the impact of the measures on all stakeholders in the city in the future.

Acknowledgements This research is completed according to the State-funded research theme "New mechanisms of horizontal collaboration for ensuring the sustainable development of transport and logistics" (registration 0119U003656 from 26-12-2019) at Department of Transport Systems and Logistics, O. M. Beketov National University of Urban Economy in Kharkiv.

References

1. BESTUFS. Good practice guide on urban freight, http://www.bestufs.net
2. CIVITAS. https://civitas.eu/TG/urban-freight-logistics
3. Sustainable Urban Transport Project (SUTP). https://www.sutp.org/
4. NOVELOG, http://novelog.eu/
5. SULPITER, https://www.interreg-central.eu
6. ENCLOSE, http://www.enclose.eu/content.php
7. DG MOVE. European Commission: Study on Urban Freight Transport (2012). https://ec.europa.eu/transport/sites/transport/files/themes/urban/studies/doc/2012-04-urban-freight-transport.pdf
8. Lindawati van Schagen J, Goh M, de Souza R (2014) Collaboration in urban logistics: motivations and barriers. Int J Urban Sci 18(2):278–290
9. Benjelloun A, Crainic T, Bigras Y (2010) Towards a taxonomy of city logistics projects. Procedia—Soc Behav Sci 2(3):6217–6228
10. Russo F, Comi A (2010) A classification of city logistics measures and connected impacts. Procedia—Soc Behav Sci 2(3):6355–6365
11. Nuzzolo A, Comi A (2014) City logistics planning: demand modelling requirements for direct effect forecasting. Procedia—Soc Behav Sci 125:239–250
12. Russo F, Comi A (2012) City characteristics and urban goods movements: a way to environmental transportation system in a sustainable city. Procedia—Soc Behav Sci 39:61–73
13. Filippi F, Nuzzolo A, Comi A, Site P (2010) Ex-ante assessment of urban freight transport policies. Procedia—Soc Behav Sci 2(3):6332–6342
14. Browne M, Allen J, Nemoto T, Patier D, Visser J (2012) Reducing social and environmental impacts of urban freight transport: a review of some major cities. Procedia—Soc Behav Sci 39:19–33
15. Papoutsis K, Nathanail E (2016) Facilitating the selection of city logistics measures through a concrete measures package: a generic approach. Transp Res Procedia 12:679–691
16. Tamagawa D, Taniguchi E, Yamada T (2010) Evaluating city logistics measures using a multi-agent model. Procedia—Soc Behav Sci 2(3):6002–6012
17. Morfoulaki M, Kotoula K, Stathacopoulos A, Mikiki F, Aifadopoulou G (2016) Evaluation of specific policy measures to promote sustainable urban logistics in small-medium sized cities: the case of Serres, Greece. Transp Res Procedia 12:667–678
18. Holguín-Veras J (2008) Necessary conditions for off-hour deliveries and the effectiveness of urban freight road pricing and alternative financial policies in competitive markets. Transp Res Part A: Policy Pract 42(2):392–413
19. Gatta V, Marcucci E (2014) Urban freight transport and policy changes: Improving decision makers' awareness via an agent-specific approach. Transp Policy 36:248–252
20. Quak H, Van Duin JHR (2010) The influence of road pricing on physical distribution in urban areas. Procedia—Soc Behav Sci 2(3):6141–6153
21. Quak H (2012) Improving urban freight transport sustainability by carriers—Best practices from The Netherlands and the EU Project CityLog. Procedia—Soc Behav Sci 39:158–171
22. Olkhova M, Roslavtsev D, Mykhalenko A (2020) Effectiveness of delivery transportation process while using electric freight vehicle. Municipal economy of cities. Series: Eng Sci Architect 3(156):2–7
23. Ant Logistics, https://ant-logistics.com
24. Olkhova M, Roslavtsev D, Matviichuk O, Mykhalenko A (2020) City delivery routes planning based on the ant colony algorithm. Sci & Technique 19(4):356–362

Simulation of Processes for Optimizing the Delivery Routes of Goods on Urban Road Networks by a Synergetic Approach

Viktor Danchuk, Vitalii Svatko, Olga Kunytska, and Yevhen Kush

1 Introduction

One of the most acute problems in modern conditions is the lagging behind the development of the road network from the real needs of the city. An increase in the level of motorization in cities leads to an intensification of traffic, a decrease in the speed of traffic, and the occurrence of traffic congestion. In turn, this significantly affects the load on the transport networks of cities, whose capacity reserves are often depleted today. Prolonged traffic jam is increasingly causing not only temporary but also economic losses.

The most promising direction for solving these problems is the development, implementation and application of intelligent transportation systems (ITS). In such conditions, special attention should be paid to the development of modern methods and models of rational organization and traffic management as a methodological basis for the creation of ITS. One of the important tasks that must be solved when using ITS is the effective management of the process of dynamic routing of freight traffic in real-time, taking into account the real dynamics of traffic flows on the urban road network (e.g. [1–4]). When solving such a problem, it is urgent to develop modern

V. Danchuk (✉) · V. Svatko · O. Kunytska
National Transport University, Mykhaila Omelianovycha—Pavlenka Street 1, 01010 Kiev, Ukraine
e-mail: vdanchuk@ukr.net

V. Svatko
e-mail: vitaliy_svatko@ukr.net

O. Kunytska
e-mail: o.kunytska@gmail.com

Y. Kush
O.M. Beketov National University of Urban Economy in Kharkiv, Bazhanova Street 18, 61002 Kharkiv, Ukraine

G. Sierpiński and E. Macioszek (eds.), *Decision Support Methods in Modern Transportation Systems and Networks*, Lecture Notes in Networks and Systems 208,
https://doi.org/10.1007/978-3-030-71771-1_12

approaches, models and methods, as well as to carry out simulation of optimization of cargo delivery processes in conditions of real dynamics of traffic flows.

2 Literature Review

To date, there are a large number of papers devoted to the development and application of various classical methods for solving problems of discrete optimization of transport routes in cities. As the analysis shows, most of the methods presented in these papers are not perfect and do not give unambiguous solutions [5–9].

In recent years, more promising for solving routing problems of different physical nature can be considered the use of evolutionary programming methods, primarily the genetic algorithm [10–12] and the method based on the effects of self-organization of ant agents [13]. Comparative analysis of these methods indicates a number of limitations when using a genetic algorithm [14, 15]. In particular, the application of a genetic algorithm leads to more accurate results in the case of solving high-dimensional routing problems. This is not always applicable to the city's transport problems. For example, the task of delivering small batches within the metropolis often contains 10–50 destinations. Accordingly, the classical ant algorithm [13] gives much better results for optimization problems with a dimension of up to 100 points [14, 15].

In addition, a large number of modifications of the ant algorithm are currently presented in the literature (see, for example, [16–18]). Here, most of the proposed modifications were mainly aimed at improving the operation of the ant algorithm, increasing the dimension and speed of computing optimization problems. As the analysis shows, modern modifications of the ant algorithm allow solving discrete optimization problems of a sufficiently large dimension. Therefore, in this chapter, for solving the problems of optimization of routes for the delivery of goods in cities, it is advisable to use, as a more universal, ant algorithm.

At the same time, it should be noted that the currently existing methods and models for optimizing the transport route mainly solve the routing problems for stationary states of the transport network, when the speeds on the sections of such networks do not change. For example, the classical method of self-organization of an ant colony [13] provides for the possibility of finding the optimal path for an urban road network in the model representation of a static graph. Here ant agents, located at the initial moment of time in the nodes of the graph, simultaneously move at equal speeds along the edges of the directed graph.

In [19], a method is proposed for optimizing the route of transportations on the basis of a modified ant algorithm, taking into account the change in the speed of traffic flows on certain sections of the road network. Here, the fundamental difference between the modification of the algorithm [19] and the currently existing modifications of the ant algorithm [16–18] is that the cyclic movement of the colony, where the ants move at constant speeds, is replaced by the asynchronous movement of each ant at a certain speed. This makes it possible to carry out simulation of route optimization in conditions of real dynamics of traffic flows on network sections, where ant agents, like vehicles analogs, move at appropriate speeds.

In [19], when simulation the processes of route optimization, the speed of traffic flow in certain sections of the road network decreased linearly without taking into account the peculiarities of the dynamics of traffic flows in the area of high density.

However, it is known that the road networks of metropolises are mainly characterized by non-stationary states, high dynamic load on its sections, significant (sharp) changes in the speed of traffic flows, the appearance of congestion, etc.

So, in particular, B. Kerner, based on the analysis of the relevant experimental data obtained on the roads of Germany, England and the United States, developed a theory of three phases of the dynamics of traffic flow (free flow, synchronized flow, wide moving jam or traffic jam) [20]. It is shown here that the occurrence of a congested traffic is explained as a first order phase transition from a free flow phase to a synchronized flow phase. This theory, unlike all previous theories, allows one to explain and predict the empirical spatiotemporal dependences of phase transitions from free flow to dense (synchronized) flow and between the resulting structures in dense traffic flow. Within the framework of this theory, it is believed that the observed effects of phase transitions, hysteresis of velocity changes during these transitions, and other nonlinear effects are due to the manifestation of synergistic effects due to the non-equilibrium non-stationary dynamics of traffic flows [21].

Here, it should be recognized that in recent years there has been a rapid development of theoretical concepts, models, methods for describing non-linear, non-equilibrium dissipative processes and phenomena occurring in systems of various physical nature, for example, in the field of phase transitions of molecular crystals, enterprise activities under the influence of external stochastic and deterministic changes in the factors of the nonlinear market environment (see, for example, [22, 23]). One of the main approaches that are used to describe such processes is a synergistic approach based on the principles of self-organization.

However, until now, the problems of developing synergistic representations of non-stationary dynamics in the area of dense traffic flow (traffic breakdown) have not received sufficient attention. Here we can cite the paper [24], where for the first time the theory of the phase transition from the free flow phase to the dense traffic flow phase was proposed in the framework of the Lorentz synergetic model. However, within the framework of this theory, the question of determining the theoretical dependence of the traffic flow speed on the vehicle density on the section of the transport network remained open [24].

The aim of this work is to carry out, within the framework of a synergetic approach, simulation of the processes of optimizing the routes of delivery of goods, taking into account the non-stationary dynamics of the traffic flow on the sections of the urban road network. Here, discrete optimization of routes is carried out using a modified self-organization algorithm of an ant colony [19]. At the same time, the analytical dependences of the change in the speed of the traffic flow on the characteristic time and the density of vehicles on the sections of the road network is determined within the framework of the Lorenz synergetic model.

3 A Synergistic Approach for Optimization of Routes on Urban Road Networks in the Conditions of Non-stationary Dynamics of Traffic Flows

3.1 Theory of the Dynamics of Traffic Flows in the Framework of the Lorentz Synergetic Model

In [20] it is experimentally shown that the occurrence of congested traffic occurs as a result of a first order phase transition from a free flow to a synchronized flow and further a wide moving jam. In [9, 20], within the framework of the theory of cellular automata, a theoretical interpretation of the dependence of speed on the traffic density is given. However, these theoretical concepts don't give satisfactory results for traffic flow dynamics in the area of congested traffic.

In [24] for the first time, within the framework of the Lorentz model, a synergetic approach is presented to describe the phase transition between different states of traffic flow dynamics. The transition between the corresponding states occurs when the vehicle density reaches a critical value. Traffic flow with increased density and an unstable homogeneous part is accompanied by a congested on the road network. In [24] it is provided a general description of the non-stationary dynamics of traffic flow for the elementary car-following model without taking into account the peculiarities of the functioning of the road network, in particular, the characteristics of vehicles that form the traffic flow in certain sections of the transport networks. This work, in contrast to [24], provides a theoretical description of the non-stationary dynamics of traffic flow in the framework of the Lorenz synergetic model, taking into account the dynamic dimensions of vehicles, which are most typical for the traffic flow, which is formed at a certain section of the network in a certain period of time.

Such vehicles in this work are considered within the framework of the concept of an effective car. In essence, an effective car is a model element of a system that describes traffic within a certain microscopic model. It has the technical and operational parameters of such a vehicle, which is the most typical for the traffic flow formed at a certain section of the network in a certain period of time. The search for such an effective car as the characteristic of the traffic flow can be carried out, for example, using the methods of cluster analysis that are sufficiently developed to date or the method of analysis of Saati hierarchies [25].

In addition, in this chapter, analytical dependences of the change in the speed of the traffic flow on the characteristic time and the traffic flow density on the sections of the road network were obtained.

So, according to [24], the transition between traffic flow modes is represented as a first order phase transition, the nature of which is determined by the distance between cars $\Delta x(t)$, which plays the role of volume or density, and the inverse delay time (acceleration/deceleration time), which corresponds to temperature. For the description, the equation of motion of the *n*-th car is used:

$$dx_n/d(t+1) = \upsilon(\Delta x_n(t)). \tag{1}$$

Here the optimal speed $\upsilon(\Delta x_n(t))$ is set, which meets the following requirements:

- $\upsilon(\Delta x_n(t))$ is a monotonically varying function;
- has an upper border;
- when the interval between cars is equal to the critical value h_c, has a point of tuning.

In the process of researching this model, a stability condition was obtained, according to which, when an insignificant interaction appears in a homogeneous flow of cars, it remains stable if the delay time corresponds to the inequality:

$$1/\tau > 2\upsilon'(h), \tag{2}$$

where $d\upsilon(\Delta x_n)/(d\Delta x_n) \mid \Delta x_n = h$.

For the most elementary car-following model, the quantities:

$$\eta = h - \Delta x,\ \Delta x = h - \eta \tag{3}$$

where η is deviation from the optimal value of the interval between vehicles; Δx is the real value of the distance between vehicles; h is the optimal value of the interval between vehicles.

$$\upsilon = \Delta\dot{x} = h/t_0 - \upsilon_0, \upsilon = \Delta\dot{x} = -\dot{\eta} \tag{4}$$

where is the rate of deviation from the optimal value of the interval between vehicles; Δx is the real value of the speed of deviation from the optimal value of the interval between vehicles.

The deviation of the interval between vehicles and the rate of its change from the corresponding optimal values of h and $h/t_0 - \upsilon_0(t_0$ is the characteristic time interval, υ_0 is the vehicle speed) play the role of the order parameter and the conjugate field, respectively. Thus, the behavior of the traffic flow is characterized by the values, υ and the acceleration/deceleration time, which is reduced to a control parameter. Let the indicated quantities be dissipative, and their relaxation at equilibrium values is described by the Debye equation. The basis of the synergistic approach is that a positive inverse relationship between the variables η and τ can lead to self-organization of the system, which is the reason for the transition between traffic flow modes. To ensure the stability of the system, we introduce a negative inverse relationship between η and. The resulting equations that determine the time dependences, $\upsilon(t)$ and $\tau(t)$ formally coincide with the Lorentz system describing a self-organizing system [2]:

$$\{\dot{\eta} = -\eta/t_\eta + \upsilon, . \tag{5}$$

$$\{\dot{\upsilon} = -\upsilon/t_\upsilon + g_\upsilon \eta \tau, \tag{6}$$

$$\{\dot{\tau} = (\tau_0 - \tau)/t_\tau - g_\tau \eta \upsilon. \tag{7}$$

Here the point means the time derivative; t_η, t_υ, t_τ are respectively the relaxation times η, υ, τ;g_υ, g_τ are additional coupling constants between dynamic variables. Equations (5)–(7) are the basis for a self-consistent description of the model of cars moving one after another. An essential feature of this model is that (6)–(7) includes nonlinear components with different signs, while Eq. (5) is linear. The latter is due to the fact that the deviation of the velocity υ is the derivative of the deviation of the interval η on time. The second component on the right side of (6) describes a positive feedback between the interval deviation and the acceleration/deceleration time, as a result of which the value of υ increases, which is the cause of the traffic jam. The kinetic equation for the value of τ differs in that the relaxation occurs not to zero, but to the final value τ_o, which determines the time required for the car to reach the characteristic speed (technical and operational characteristic of the vehicle). The stationary value of the deviation of the interval η has the form:

$$\begin{aligned} \eta_e^m &= \eta_{00}\{1 \pm [1 + \eta_0^2 \eta_{00}^{-4}(\tau_0 - \tau_c)]^{1/2}\}^{1/2} 2\eta_{00}^2 = \\ &= (\tau_0 - 1) - \tau_c \eta_0^2,\ \tau_c = 1 + k. \end{aligned} \tag{8}$$

The upper sign on the right side of Eq. (8) corresponds to the value η^m for an unstable state, and the lower sign corresponds to a stable state η_e. Corresponding value of stationary acceleration/deceleration time:

$$\tau^m = \frac{1 + \eta_{00}^2 + \sqrt{(1 + \eta_{00}^2)^2 - (1 - \eta_0^2)^2 \tau_0}}{1 - \eta_0^2}, \tag{9}$$

which gradually increases from value:

$$\tau_m = 1 + \eta_0 \sqrt{k/(1 - \eta_0^2)} \tag{10}$$

for

$$\tau_0 = \tau_{c0} = (1 - \eta_0^2)\tau_m^2 \tag{11}$$

to the limit value $\tau_c = 1 + k$ for $\tau_0 = \tau_c$.

According to (4), it is easy to find the vehicle speed υ in a synchronized mode, when there is an interaction between vehicles, which leads to the formation of a traffic jam at last. In this case, the change for value of the speed is due to the manifestation of synergistic effects, which are described by the system of Eqs. (5–7):

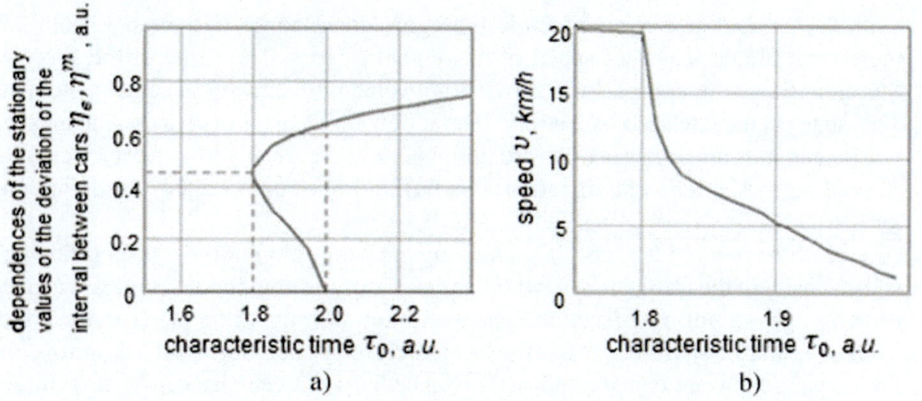

Fig. 1 Dependences of the stationary values of the deviation of the interval between cars $\tilde{n}_e, n_i$ (**a**) and the vehicle speed υ (**b**) on the characteristic time τ_0

$$\upsilon = \eta_e^m = \frac{1}{\sqrt[4]{\eta_{00}^2 \mp \sqrt{\eta_{00}^4 + \eta_0^2(\tau_0 - \tau_c)}}}(1 \mp \frac{\eta_0^2 - \eta_{00}^2}{\sqrt{\eta_{00}^4 + \eta_0^2(\tau_0 - \tau_c)}}). \tag{12}$$

The upper sign on the right side of Eq. (12) corresponds to the value for an unstable state, and the lower sign corresponds to a stable state.

Based on the calculations obtained, we construct the dependence of the stationary value of the deviation from the optimal value of the interval between vehicles η_θ^m and the speed from the value τ_0, which determines the time required for the car to reach the characteristic speed. The η_e^m, υ, τ_0, values are given in relative units.

As shown in Fig. 1a, with a slow increase in the parameter τ_0 in the range from 0 to $\tau_c\mathbf{0} = 1.8$, the occurrence of a traffic jam is impossible. At the point $\mathbf{J}_2 = \mathbf{o}_2$, the deviation of the interval $\eta_{e_}$e _e causes a jump to the value $\sqrt{2}\eta_{00}$, and then gradually increases according to (8). As the parameter τ_0 decreases, the value η_e continuously decreases to the point $^{0o_1} = \mathbf{o}_2, \eta_e = \eta_{00}$, and then drops sharply to zero.

As shown in Fig. 1b, a slow increase in the parameter τ_0 in the interval from 0 to τ_{C0}, equal to 1.8, traffic speed practically does not change, as a result, a traffic jam cannot occur. In this interval, vehicles move in free traffic mode. Here, the speed of movement of vehicles is limited only by the rules of the road, the technical characteristics of the vehicle and the qualifications of the driver.

In the interval from τ_{C0} to τ_C, which in the figure corresponds to the range from 1.8 to 2.0, there is a sharp decrease in the velocity value, which rapidly approaches zero. This traffic flow state is characterized as a synchronized flow (according to Kerner's classification). This state is characterized by a sharp decrease in traffic flow speed. In a synchronized traffic flow, the speed of vehicles is less than the minimum possible speed of vehicles in free traffic. In this area, there is a significant interaction between vehicles in the traffic, the speed of the vehicle depends on the total speed of the traffic and cannot be greater than it.

In the interval above τ_C, which in the figure corresponds to the range from 2.0, there is an almost constant speed of movement of cars. The value of the speed in this area is actually approaching zero, the movement of vehicles is actually stopped. This state is characterized by a strong interaction between all road users, maneuvers, rearrangements and other actions are possible only if agreed with other participants. According to Kerner's classification, this traffic flow state is called a wide moving jam.

One of the main indicators characterizing the congestion of the transport network is the density of the traffic flow ρ, which is defined as the number of vehicles at a given time on a given unit road section. The maximum density value ρ_{max} corresponds to the number of stationary vehicles located close to each other. Based on this, the relationship between density and real distance between vehicles can be determined using the following formula:

$$\rho = \frac{1}{\Delta x} \tag{13}$$

where ρ is the load density of the transport network (or traffic flow density); Δx is real distances between vehicles.

Since, according to (3), $\Delta x = h - r$ we have:

$$\rho = \frac{1}{h - \eta} \tag{14}$$

In addition, considering the model of an effective car as a characteristic of a vehicle of a real traffic flow, it is possible to describe the dynamics of phase transitions of a traffic flow in Kerner's theory within the framework of the synergetic approach developed in this work.

Then, using formulas (12), (14), it is possible to construct the dependence of the speed of movement of the traffic flow υ on the indicators of the congestion of the road transport network.

As shown in Fig. 2, the traffic flow rate remains invariably high until a certain value of the traffic flow density is reached in a certain section of the transport network (in our case, to the value). At this point, a phase transition occurs from the free mode of movement of vehicles to the synchronized mode. Here, the speed of the traffic flow begins to decrease sharply with increasing traffic density. When the critical density value is reached (in our case, the value), the traffic flow velocity value approaches zero and then practically does not change. This state corresponds to the state of movement of a wide cluster (wide moving jam by Kerner's definition). Under certain conditions, the speed can drop to zero, which corresponds to the state of a traffic jam.

So, here, within the framework of the Lorentz synergetic model, an interpretation of the Kerner's phase diagram of the transition from free to congestion traffic flow is obtained. The presence, according to [20], of three phases in the traffic flow is theoretically confirmed: free movement, synchronized movement and wide moving jam. Analytical dependences of the speed of movement υ on the density of the traffic

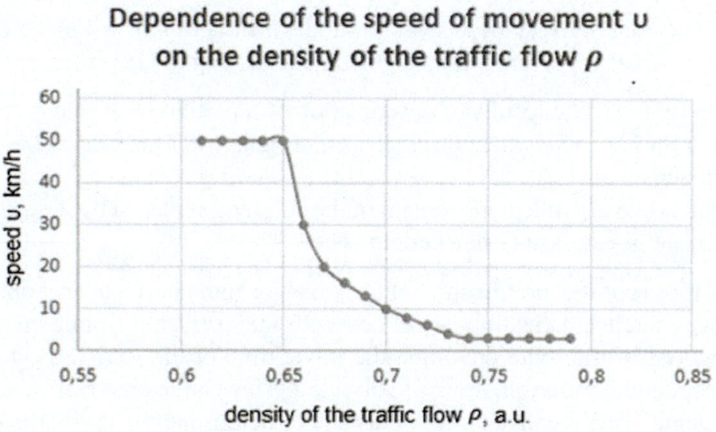

Fig. 2 Dependence of the speed of movement υ on the density of the traffic flow ρ on the section of the transport network

flow ρ are obtained, which indicate the presence of clear boundaries of the transition between the phases of the dynamics of the traffic flow.

3.2 *Self-organization Ant Colony Method for Optimizing the Route of Delivery of Goods*

According to [19], the urban road network will be represented as a bidirectional weighted graph. In the nodes of such a graph, there are points of delivery of goods (warehouses, supermarkets, etc.) from the sender's warehouse using a vehicle moving along the city road network. The physical meaning of the weights of the edges connecting a certain pair of nodes in the graph depends on the nature of the problem being solved. These can be distances between points, average speed or average time, the cost of travel by a vehicle on certain sections of the network. Optimization of routes of movement of a vehicle is carried out within the framework of the traveling salesman problem using an ant colony self-organization algorithm [13]. This problem is formulated as the problem of finding a closed route that is minimal in a certain parameter over all vertices visited by a traveling salesman, without repetitions on a certain weighted graph with m vertices and n edges. Here, optimization parameters can be route length, average time, value of travel along the entire route. For convenience, we will assume that the departure and return of the vehicle can occur from any node of the graph.

To solve routing problems, taking into account the real behavior of the traffic flow (congestion, road repair, emergency situations, etc.), we will consider a dynamic bidirectional weighted graph. The weights of the edges in this graph are determined by the average travel time of the vehicle in the traffic flow at certain sections of the network between delivery points (graph nodes). In addition, we make appropriate changes to the classical ant colony algorithm according to [19], namely:

- the ability to fix the results of optimization of a partially traveled path for calculating a further route when the edge weight (length) of the graph changes during the movement;
- the synchronous cyclical movement of the colony is replaced by the asynchronous movement of each ant with a certain speed.

Thus, the proposed modification of the ant algorithm takes into account additional elements of intellectualization associated with the exclusion from consideration of those sections of the route on which the travel time of the vehicle is, according to the corresponding criterion, unjustifiably long. This can lead to restructuring of the optimal route. This means that it is possible to determine the optimal route for the real state of the traffic flow (change in speed, traffic congestion, etc.).

Such a modified method of the ant algorithm for optimizing the delivery route of goods taking into account the dynamics of the traffic flow is applied under the following assumptions:

- at all sections of the network, the movement of the dedicated vehicle is carried out within the framework of a two-lane two-way traffic flow;
- in each set of n_j network sections, which corresponds to the j-th edge of the graph, there will always be alternative travel options;
- changes in average travel time and average travel speed mainly depend on changes in traffic flow dynamics, modes that do not include stops or delays due to traffic lights signals.

Each route between the point of departure and the point of delivery of goods within the transport network has a certain number of alternatives. This means that it is possible to deliver cargo from one point to another by several routes, which differ only in the length and average speed of the traffic flow on them. In our case, we will assume that the average travel time along the n-th edge between two nodes of the graph is found as follows:

$$t_n = \frac{l_n}{\upsilon_n}, \tag{15}$$

where t_n is the average time spent on passing the n-th edge, hours;

l_n is length of the n-th edge, km;

υ_n is average speed of traffic flow on the n-th edge, km/h.

Since the value of the real speed of traffic flows in each section of the road network can change, then the value for the time of travel along corresponding sections of the city road network is also a variable. Therefore, when building the time-optimal route for the delivery of goods, information about the real state of the network at the time of building the route is taken into account. At the same time, of course, for some limit values of the speed of movement of the traffic flow in certain sections of the network, it is possible of rebuilding of the time-optimal route for delivery of the goods to the destinations.

A synergistic method for optimizing the route of delivery of goods on an urban street-road network, taking into account the non-stationary dynamics of traffic flow on its sections, which is proposed in this work, is as follows. First, the parameters of the point of departure and points of delivery of goods on the corresponding urban road network or on its fragment are determined. Further, experimental observations of the dependence of the traffic flow density on time are carried out on the network sections for the selected time interval. Thereafter, using the analytical dependences of the traffic flow speed on the density (12)–(14) (see Fig. 2), obtained within the framework of the Lorentz synergistic model (see Sect. 3.1), we determine the speed of the traffic flow in these sections of the network at certain points in time. After that, within the framework of the modified ant algorithm (see Sect. 3.2.) simulation of route optimization processes is carried out depending on the speed of traffic flows on sections of the road network.

4 Case Study: Application of the Method for Kyiv Road Network. Simulation Results and Discussion

The Kyiv city was chosen to simulate the processes of optimizing the route of delivery of goods. This choice is due to several factors:

- a large number of outlets (customers, shops, shopping centers, etc.);
- presence of an branched road transport network;
- high intensity and density of traffic flows on most sections of the road network;
- high probability of congestion in areas with high traffic density;
- availability of systems of fixing traffic flows (video surveillance cameras in real time) and open access to them;
- ability to compare simulation results with experimental data.

20 points of delivery of goods were chosen, located in several districts of the Kiev city. These are Darnitsky, Desnyansky and Pechersky districts. Information about outlets and addresses of their location are given in Table 1.

We will assume that the sequence of passing the indicated points is determined by the time of their traversing. The average travel time at each section of the network was determined according to (15). The construction of the optimal route for bypassing all 20 delivery points, taking into account the real dynamics of traffic flows at a certain point in time, was carried out according to the method described in 3.2.

The initial data for modeling were the distances between points and the time of traffic flow at each section of the network, which was obtained for free movement on these sections. At the same time, the speeds on the sections were chosen as maximum, which corresponded to the value of road signs of speed limits (50 km/h). These data are shown in Table 2.

Table 1 Names and location of outlets for the selected fragment of the Kyiv urban road network

Number of outlet	Outlet name	Location
1	ATB-Market	Malyshko street 25/1
2	Silpo	Darnytskyi Blvd. 8A
3	Furshet	Bratislava street 14B
4	ATB-Market	Sholom Aleichem street, 4
5	Bdzhilka	Sholom Aleichem street 10A
6	ATB-Market	Academician Kurchatov street 8a
7	ATB-Market	Sholom Aleichem street, 17
8	Fora	Academician Kurchatov street 19
9	ATB-Market	Forest Avenue 28
10	Silpo	Forest Avenue 39
11	Novus	Brovarsky Avenue 17
12	Fora	Perov Blvd. 20
13	ATB-Market	Rainbow street 49
14	Furshet	Petro Vershigora street 1
15	Silpo	Rainbow street 8
16	Novus	Friendship of Peoples Blvd. 16A
17	Silpo	M. Dragomirov street 16
18	FOZZY	S. Bandera Avenue 23
19	ASHAN	S. Banderai Avenue 15-A
20	METRO	S. Bandera Avenue 26-V

The optimization criterion is the time of the cargo delivery route to all selected retail outlets. Using a modified ant algorithm (see 3.2), the time-optimal route for the delivery of goods to 20 outlets was found for the initial simulation data. This optimal route has the following form: 10–9–5–7–4–6–3–2–1–12–15–14–19–20–18–16–17–11–13–8–10 (numbering outlets corresponds to the numbering given in Table 1). In this case, the delivery time of goods is 86.2 min, and the length of the route is 71.65 km.

To carry out simulation of the processes of building the optimal route in conditions close to the real state on the urban road network, three network sections were selected, on which experimental observations of the dependence of the traffic flow density on time during the day were carried out. These are such sections (see Fig. 3):

- North bridge in both directions (network Sect. 14–19)
- Druzhby Narodov metro station–Paton Bridge (network Sect. 17–11);

Table 2 Initial values of delivery speeds and distances between outlets in the selected fragment of the Kyiv urban road network

Points	Values of distances between points (km) Initial values of delivery speed (km/h)																			
between objects	1	2	3	4	5	6	7	8	9	10	11	12	13	14	15	16	17	18	19	20
1		50	50	50	50	50	50	50	50	50	50	50	50	50	50	50	50	50	50	50
2	1.3		50	50	50	50	50	50	50	50	50	50	50	50	50	50	50	50	50	50
3	2	3.3		50	50	50	50	50	50	50	50	50	50	50	50	50	50	50	50	50
4	2.3	1.8	0.95		50	50	50	50	50	50	50	50	50	50	50	50	50	50	50	50
5	2.9	2.3	1.5	0.6		50	50	50	50	50	50	50	50	50	50	50	50	50	50	50
6	2.7	2.3	1.5	1.8	3.1		50	50	50	50	50	50	50	50	50	50	50	50	50	50
7	3.8	3.2	2.4	1.5	1.9	2.3		50	50	50	50	50	50	50	50	50	50	50	50	50
8	4	3.4	2.5	1.6	2.6	1.2	1.2		50	50	50	50	50	50	50	50	50	50	50	50
9	4.2	3.9	2.9	2.1	2.7	1.7	1.3	1.3		50	50	50	50	50	50	50	50	50	50	50
10	4.9	4.8	3.5	2.5	2.9	2.2	1.6	1.5	0.55		50	50	50	50	50	50	50	50	50	50
11	2.5	2.8	5.5	5	6.2	5.7	5	5.5	6	6.6		50	50	50	50	50	50	50	50	50
12	3.8	2.7	3.4	3.7	5	4.4	3.7	4.2	4.7	5.3	4.4		50	50	50	50	50	50	50	50
13	5.1	4	4.7	5	6.3	4.3	5	5.1	4.6	5.2	5.8	1.4		50	50	50	50	50	50	50
14	6.6	5.5	6.1	6.4	7.8	5.8	6.4	6.6	6.2	6.7	7.3	2.8	2.4		50	50	50	50	50	50
15	5.9	4.8	5.5	5.8	7.1	5.2	5.8	5.9	5.1	6	6.6	2.2	1.5	1.1		50	50	50	50	50
16	13	13.3	14.4	14.9	16.2	15.5	14.8	15.3	15.8	16.4	11.3	14.5	15.2	17.2	16.3		50	50	50	50
17	10.7	11.1	11.9	12.3	13.7	13	12.3	12.8	13.3	13.9	8.4	12.2	12.9	15	14	1.3		50	50	50
18	12.7	11.5	11.9	12.3	13.6	11	12.2	11.7	11.3	11.8	13.3	8.8	8.5	7.4	8	14.6	15.1		50	50
19	13.2	12.5	12.5	12.8	14.1	11.5	12.8	12.3	11.9	12.4	13.8	9.4	9	7.9	8.6	15.2	15.6	0.55		50

(continued)

Table 2 (continued)

Points between objects	Values of distances between points (km) Initial values of delivery speed (km/h)																			
	1	2	3	4	5	6	7	8	9	10	11	12	13	14	15	16	17	18	19	20
1		50	50	50	50	50	50	50	50	50	50	50	50	50	50	50	50	50	50	50
20	13.3	11.7	12.1	12.4	13.7	11.2	12.4	11.9	11.5	12	13.3	9	8.6	7.5	8.2	14.8	15.2	2.9	2.5	

[a]Below the main diagonal of the matrix, there are distances between any two outlets, and above the main diagonal of the matrix, there is initial speed between them. The initial speed at each section with free movement of vehicles in the traffic flow is 50 km/h and is limited only by traffic rules

Fig. 3 The investigated sections of the Kyiv urban road network: **a** metro Druzhby Narodov—Paton Bridge; **b** North Bridge; **c** Bratislavskaya Street—Lesnaya Avenue

- intersection of Bratislavskaya Street and Lesnaya Avenue in both directions (network Sect. 13–8).

The choice of these sections for a simulation study is due to their high workload, as well as the frequent occurrence of traffic jams and traffic complications during the day, which leads to a significant decrease in the speed of traffic flows on them.

Each of the selected sections of the network has certain features in its characteristics. So, the section of the network Druzhby Narodov metro network—the Paton Bridge in both directions has 3 lanes of vehicle traffic and one lane of public transport. In addition, traffic light regulation is provided in the specified area. On the section of the network North Bridge network in both directions, there are three lanes for vehicle traffic and one lane for public transport in each direction. There are no additional restrictions on the selected section as well as there is no traffic light regulation. The section of the network at the intersection of Bratislavskaya Street and Lesnaya Avenue has three lanes for traffic flows of cars and one lane for public transport in each direction. Traffic light regulation is provided in the specified section.

During the study, for 5 days with an interval of 1 min, daily observations were made of the number of vehicles filling the corresponding sections of the network at certain times of the day. Average values of the number of cars for certain points in time were determined by five measures. The data obtained were further used to calculate such indicators of traffic flow as average intensity and average density. Further, according to (12)–(14) (see Fig. 2), for certain points in time, the average speeds of traffic flows were determined depending on the indicators of the average density in the indicated sections of the network. The relative measurement error for all indicated indicators did not exceed 7–10%. At all other sections of the network the speeds of traffic flows were considered equal to 50 km/h.

The resulting dependences of the average density of traffic flows on the time of day at the studied sections of the network during the day are shown in Figs. 4, 5 and 6.

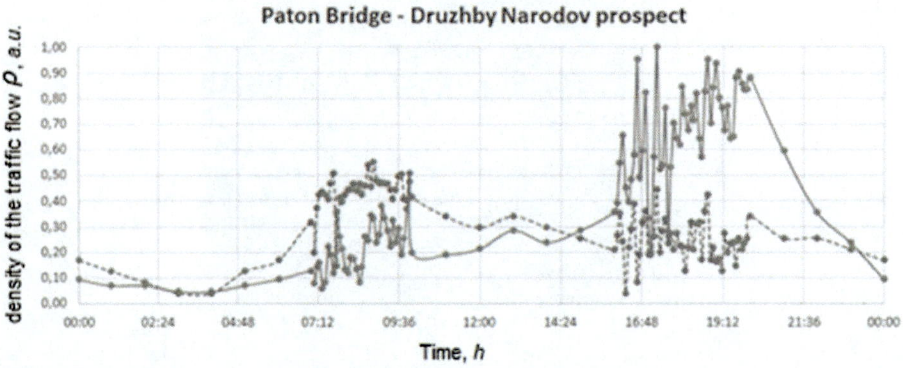

Fig. 4 Dependence of the average traffic density on time during the day on the section of the Druzhby Narodov prospect—Paton bridge (the solid line shows the density in the forward direction, the dashed one—in the opposite direction)

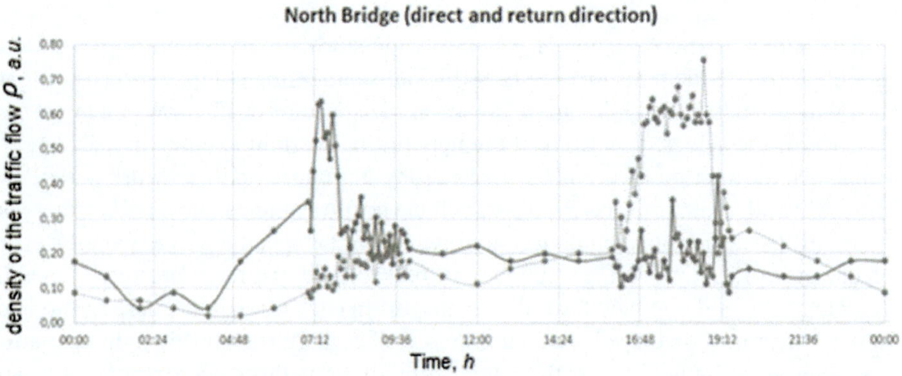

Fig. 5 Dependence of the average traffic density on time during the day on the section of the North bridge (the solid line shows the density in the forward direction, the dashed one—in the opposite direction)

Fig. 6 Dependence of the average traffic density on time during the day on the section of the intersection of Bratislavskaya Street and Lesnaya Avenue

Further, on the basis of the obtained data on the average density of traffic flows in the studied sections of the network at certain time during the day, the corresponding speeds of traffic flows in these sections were determined according to (12)–(14) (see Fig. 2).

This made it possible, using the modified ant algorithm (see 3.2), to carry out simulation in real time of the processes of building the optimal route for the delivery of goods to outlets on a selected fragment of the Kyiv urban road network taking into account the real dynamics of traffic flows on three model sections of the network during the observation day. At the same time, in these sections, the speed varied in the range from 3 to 50 km/h, depending on the value of the traffic flow density on them. In all other sections of the road network fragment, the speed remained constant and equal to 50 km/h.

During the simulation, a number of effects were found (see Table 3). These effects are associated with the rebuilding of optimal routes when average speeds of vehicles on the model sections of the network decrease to certain boundary values corresponding to certain modes of traffic flow.

So, during the day in the ranges from 00:00 to 7:05; from 8:50 to 9:15; from 10:00 to 17:10; 17:20; from 22:00 to 24:00, the density of traffic flows in the selected sections is less than 0.6, which corresponds to the free movement of traffic. The optimal route in this case is: 10–9–5–7–4–6–3–2–1–12–15–14–19–20–18–16–17–11–13–8–10. Travel time the route is 86.2 min, and its length is 71.65 km. The obtained parameters for these states of the urban road network, which is understandable, completely coincide with the initial data given above. We observe the same optimal route at 07:10 and 21:00, although here at Sect. 17–11 (Druzhby Narodov metro station—Paton Bridge), cars move in a synchronized flow (see Table 3). This only leads to a slight increase in the route time (87.8 min) due to a decrease in speed in this section to 43 km/h.

Table 3 Results of simulation of processes of building of the optimal route of delivery of goods depending on real dynamics of transport flows on sections of the Kyiv urban road network (values of speed υ and density ρ of the traffic flow on model sections of a network which correspond to a dense state of loading of a transport network are marked by a gray background)

Time of day h	14–19		17–11		13–8		Route	Route time, min	Route length, km
	υ, km/h	ρ, (a.u.)	υ, km/h	ρ, (a.u.)	υ, km/h	ρ, (a.u.)			
00:00–07:05	50	<0.6	50	<0.6	50	<0.6	10–9–5–7–4–6–3–2–1–12–15–14–19–20–18–16–17–11–13–8–10	86.2	71.65
07:10	50	0.6	43	0.66	50	0.5	10–9–5–7–4–6–3–2–1–12–15–14–19–20–18–16–17–11–13–8–10	87.8	71.65
07:15	30	0.67	23	0.68	50	0.63	3–6–9–7–10–12–13–15–14–2–5–1–11–16–17–18–19–20–8–4–3	88.7	73.8
07:20–08:05	18	>0.69	13	>0.7	13	>0.7	13–6–10–5–4–3–8–9–7–2–11–12–14–15–20–18–19–16–17–1–13	89.2	74.3
08:10–08:45	50	<0.6	50	<0.6	13	>0.7	8–2–4–10–7–5–12–13–15–14–18–19–20–16–17–9–11–1–3–6–8	88.1	73.25
08:50–09:15	50	<0.6	50	<0.6	50	<0.6	10–9–5–7–4–6–3–2–1–12–15–14–19–20–18–16–17–11–13–8–10	86.2	71.65
09:20	50	<0.6	50	<0.6	23	0.68	1–2–5–7–13–12–14–4–8–6–3–9–10–15–20–19–18–16–17–11–1	87.7	72.9
09:25	50	<0.6	50	<0.6	3	0.79	8–2–4–10–7–5–12–13–15–14–18–19–20–16–17–9–11–1–3–6–8	88.1	73.25
09:30–09:40	50	<0.6	50	<0.6	15	>0.7	1–2–5–7–13–12–14–4–8–6–3–9–10–15–20–19–18–16–17–11–1	87.7	72.9
09:45–09:55	50	<0.6	50	<0.6	3	0.8	8–2–4–10–7–5–12–13–15–14–18–19–20–16–17–9–11–1–3–6–8	88.1	73.25
10:00–17:10	50	<0.6	50	<0.6	50	<0.6	10–9–5–7–4–6–3–2–1–12–15–14–19–20–18–16–17–11–13–8–10	86.2	71.65
17:15	50	<0.6	3	0.9	50	<0.6	8–2–4–10–7–5–12–13–15–14–18–19–20–16–17–9–11–1–3–6–8	88.1	73.25
17:20	50	<0.6	50	<0.6	50	<0.6	10–9–5–7–4–6–3–2–1–12–15–14–19–20–18–16–17–11–13–8–10	86.2	71.65
17:25	50	<0.6	50	<0.6	3	0.79	8–2–4–10–7–5–12–13–15–14–18–19–20–16–17–9–11–1–3–6–8	88.1	73.25
17:30–20:00	50	<0.6	3	<0.73	50	<0.6	8–2–4–10–7–5–12–13–15–14–18–19–20–16–17–9–11–1–3–6–8	88.1	73.25
21:00	50	<0.6	43	0.66	50	<0.6	10–9–5–7–4–6–3–2–1–12–15–14–19–20–18–16–17–11–13–8–10	87.8	71.65
22:00-24:00	50	<0.6	50	<0.6	50	<0.6	10–9–5–7–4–6–3–2–1–12–15–14–19–20–18–16–17–11–13–8–10	86.2	71.65

The situation changes fundamentally at other time intervals. Here, the traffic intensity on certain model sections of the studied network increases significantly. In some cases, the movement of traffic flows is carried out not only in a synchronized mode, but also in the mode of movement of a wide cluster, traffic jam occurs (see Fig. 2, Table 3). This leads to a significant decrease in speed in these sections of the network and, accordingly, to a rebuilding of the optimal route as a whole. At the same time, some or all of the model sections of network are excluded from the new configurations of the optimal route: 14–19 (North Bridge), 17–11 (Druzhby Narodov metro station—Paton Bridge), 13–8 (intersection of Bratislavskaya Street and Lesnaya Avenue) (see Table 3).

So, for example, in the time interval from 09:30 to 9:40 am, the traffic density in Sect. 13–8 of the network becomes more than 0.7, which corresponds to the mode of movement of a wide cluster in this section, with free movement of cars in other sections of the network (see Table 3). Calculations using the proposed method (see Sect. 3) show that here the optimal route for the delivery of goods is rebuilding so that two model network Sects. 14–19 and 13–8 are removed from it. In the next observation time interval from 9:45 to 9:55, the traffic density in Sect. 13–8 continues to increase, bringing the state of traffic in this section closer to the traffic jam. This leads to a rebuilding of the optimal route from which all three model Sects. 14–19, 13–8 and 17–11 are excluded (see Table 3). The optimal route has the following configuration: 8–2–4–10–7–5–12–13–15–14–18–19–20–16–17–9–11–13–6–8; the time of passing the optimal route is 88.1 min, the length of the optimal route is 73.25 km. A similar situation takes place in other periods of time for monitoring the dynamics of traffic flow on the road network of the Kyiv city (see Table 3).

The dynamics of changes in the time and length of the optimal route during the day are shown in Figs. 7 and 8.

So, the analysis of the results of simulation studies gives reason to believe that the proposed synergistic approach makes it possible to adequately optimize cargo

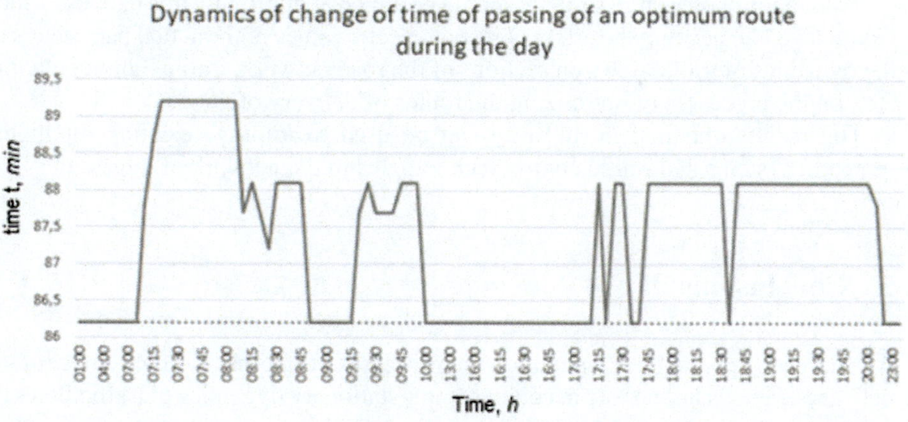

Fig. 7 Dynamics of change of time of passing of an optimum route during the day

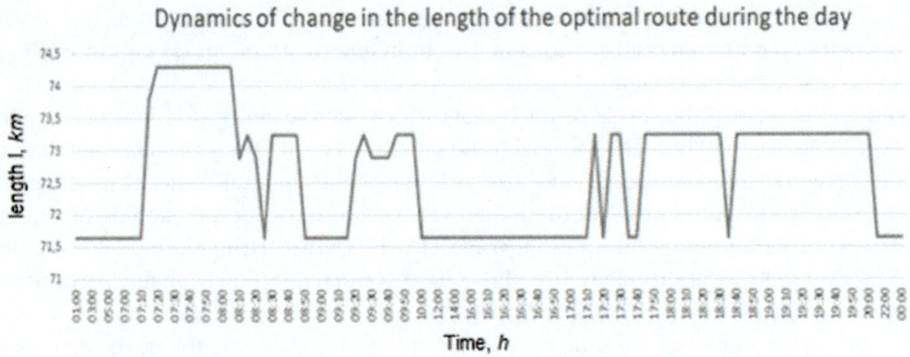

Fig. 8 Dynamics of change in the length of the optimal route during the day

delivery routes in real time under conditions of non-stationary dynamics of traffic flows in the network of large cities. In the future, it can find its application in ITS, for example, in solving problems of dynamic routing of vehicles with using traffic prediction information [26]. In [26], the main idea of the algorithm is to use real-time traffic information as well as predicted traffic information provided by a central routing controller. In our case, in real time, it is necessary to use information about the traffic density in the network sections. This information can be obtained by using the vehicle's on-board communication equipment with other vehicles and (or) transport systems infrastructure (V2V, V2I or I2V). Further, based on the real traffic information obtained at certain points in time within the framework of the synergetic approach for description of traffic flow dynamics, which was developed in this chapter (see 3 and 4), simulate alternative routes are carried out, from which the most optimal ones is determined. Accordingly, the obtained information about the optimal routes depending on the initial traffic conditions at certain points in time is provided to the user by the central routing controller.

Thus, a further research strategy should be aimed at improving this method, which would take into account the effect of instantaneous values of the actual parameters of the dynamics of traffic flows on sections of the road network, traffic light regulation, etc., on the processes of optimizing the routes of delivery of goods.

The results obtained in this way can be used to improve existing intelligent transport systems and traffic control technologies in the network of large cities.

5 Conclusions

A synergetic approach for simulation the processes of optimization of routes for the delivery of goods taking into account the non-stationary dynamics of traffic flows on sections of the urban road network is proposed. Within the framework of synergetic Lorentz model, the analytical dependences of the change in the speed of the traffic flow on the characteristic time of the change in the dynamics of the traffic flow and the density of the traffic flow in the network sections are determined. Further, on the basis of the obtained dependence of the traffic flow speed on its density by using a modified self-organization algorithm of the ant colony, the route on the network is optimized. Here the urban road network is represented as a bidirectional weighted graph. The main element of the modification is that it implements the possibility of asynchronous movement of the each ant colony agent at a certain speed. In addition, it is also possible to fix the results of optimization of a partially traveled path for calculating a further route when the edge weight (length) of the graph changes during the movement. This allows you to management the route optimization process, taking into account the dynamic state of the network, which depends on the speed of movement of vehicles on certain sections of the network. These changes in speed can be caused by an increase in the congestion of network sections, the occurrence of traffic jams, emergency situations, etc.

To test the proposed method, simulation of route optimization processes was carried out within the framework of the traveling salesman problem, taking into account the real dynamics of traffic flows on the example of the road network of Kyiv city. Here, when conducting research on some selected sections of the network, the average density of traffic flows at certain points in time during the day was experimentally determined. Based on the simulation results, a number of effects were identified. These effects are associated with the rebuilding of optimal routes when the average speeds of vehicles on the model sections of the network decrease to certain boundary values corresponding to certain modes of traffic flow.

The chapter discusses the possibility of using the proposed method in intelligent transportation systems. In particular, this method can be used to solve dynamic vehicle routing problems using information about traffic prediction.

References

1. Kiciński M, Witort P, Merkisz-Guranowska A (2019) Multiple criteria optimization for supply chains—analysis of case study. In: Sierpiński G (ed) Integration as solution for advanced smart urban transport systems, vol 844. Advances in intelligent systems and computing. Springer, Cham, pp 150–160
2. Galińska B (2020) MCDM as the tool of intelligent decision making in transport. case study analysis. In: Sierpiński G (ed) Smart and green solutions for transport systems. Advances in intelligent systems and computing, vol 1091. Springer, pp 67–79
3. Macioszek E, Sierpiński G, Staniek M (2017) Analysis of trends in development of freight transport logistics using the example of Silesian Province (Poland) – a case study. Transp Res Procedia 27:388–395
4. Macioszek E, Sierpiński G (2020) Charging stations for electric vehicles—Current situation in Poland. In: Mikulski J (ed) Research and the future of telematics, communications in computer and information science, vol 1289. Springer, Heidelberg, pp 124–137
5. Lukinskiy V, Dobromirov V (2016) Methods of evaluating transportation and logistics operations in supply chains. Transp Telecommun 17:55–59
6. Knight H (2014) New algorithm can dramatically streamline solutions to the 'max flow' problem, MIT News, 21–26

7. Balasescu S, Balasescu M (2014) Optimization methods for supply chain activities. Bull Transilvania Univ of Brasov Series V: Econ Sci 7(56):9–16
8. Qingyou Y, Zhang Q (2015) The optimization of transportation costs in logistics enterprises with time-window constraints. Discrete Dyn Nat Soc 2015:10–15
9. Prokudin G, Chupaylenko O, Dudnik O, Oliskevych M (2019) Development of vehicle speed forecasting method for intelligent highway transport system. Eastern-Eur J Enterprise Technol 4(3):6–14
10. Glover F, Kochenberger G (2003) Handbook of metaheuristics. In: International series in operations research & management science, vol 57, p 570
11. McCall J (2005) Genetic algorithms for modelling and optimization. J Comput Appl Math 184:205–222
12. Rejer I, Lorenz K (2015) Classic genetic algorithm vs. Genetic algorithm with aggressive mutation for feature selection for a brain-computer interface. Przegląd Elektrotechniczny 91(2):98–102
13. Dorigo M, Gambardella LM (1997) Ant colonies for the travelling salesman problem. BioSystems 43(2):73–81
14. Chandekar N, Jayachandran Pillai M (2017) A comparative study of GA and ACO for solving travelling salesman problem. Int J Mech Prod Eng 5(11):34–37
15. Ali H, Haris M, Hadi F, Ahmadullah S, Shah Y (2016) Solving traveling salesman problem through optimization techniques using genetic algorithm and ant colony optimization. J Appl Environ Biol Sci 6(4S):55–62
16. Seyedali M (2018) Ant Colony optimisation. Evolutionary algorithms and neural networks, pp 33–42
17. Abdullah Haslina, Ramli Rizauddin (2017) Dzuraidah Abd Wahab: tool path length optimisation of contour parallel milling based on modified ant colony optimisation. Int J Adv Manuf Technol 92:1263–1276
18. Chiranjit C, Mahapatra GS, Rajat Kumar P (2017) A modified ant colony optimisation based approach to solve sub-tour constant travelling salesman problem. Int J Math Oper Res 11:3
19. Danchuk V, Bakulich O, Svatko V (2019) Building optimal routes for cargo delivery in megacities. Transp Telecommun 20(2):142–152
20. Kerner BS (2009) Introduction to modern traffic flow theory and control. Springer, Berlin
21. Haken H (2004) Synergetics. Introduction and advanced topics. Springer, Berlin, Heidelberg
22. Puchkovska GO, Makarenko SP, Danchuk VD, Kravchuk AP, Baran J, Kotelnikova EN, Filatov SK (2002) Dynamics of molecules and phase transitions in the crystals of pure and binary mixtures of n-paraffins. J Mol Struct 614(1):159–166
23. Danchuk MV, Kravchuk AP (2013) Features of value-at-risk methodology application for business risks estimation under nonlinear dynamics of economic development. Actual Probl Econ 148(10):207–213
24. Olemskoi A, Khomenko A (2001) Synergetic theory for jamming transition in traffic flow. Phys Rev E 63(3)
25. Saati T (1989) Priniatie resheniy. Metod analiza ierarhiy. Moskva, Radio i sviaz
26. Kwangsoo K, Minseok K, Jaegeun P, Yongsoon E (2016) Dynamic vehicular route guidance using traffic prediction information. Hindawi Publishing Corporation—Mobile Information Systems, pp 1–11

Methods of Analytical Modeling the Process of Freight Transportation Management in the Regional Transport Complex

Oleg Chislov, Vyacheslav Zadorozhniy, Taras Bogachev, Alexandra Kravets, Victor Bogachev, and Irina Egorova

1 Introduction

Modern conditions for the organization and implementation of cargo transportation by rail involve the interaction of several participating parties (carrier, owner of rolling stock, cargo owner, operator, etc.), operating in a competitive environment. When carrying out export-import transportations, multimodal transport schemes are often used, in which road and water transport entities, in particular sea ports, also participate. It is necessary to keep in mind the increasing volume of traffic through seaports, which is largely due to the raw material nature of the Russian economy, as well as the high potential of international transport corridors that transit cargo flows. Raw materials exported from Russia have a mass, volume and relatively low cost, which determines the greatest use of rail and water transport in their transportation.

A number of works are devoted to the study of improving the efficiency of the transportation process, the structure of which includes transshipment in seaports

O. Chislov · V. Zadorozhniy (✉) · A. Kravets · V. Bogachev · I. Egorova
Rostov State Transport University, Rostov-on-Don, Russia
e-mail: zadorozniy91@mail.ru

O. Chislov
e-mail: o_chislov@mail.ru

A. Kravets
e-mail: kravec_as@mail.ru

V. Bogachev
e-mail: bogachev-va@yandex.ru

I. Egorova
e-mail: egorova_irina01@mail.ru

T. Bogachev
Rostov State University of Economics, Rostov-on-Don, Russia
e-mail: bogachev73@yandex.ru

G. Sierpiński and E. Macioszek (eds.), *Decision Support Methods in Modern Transportation Systems and Networks*, Lecture Notes in Networks and Systems 208,
https://doi.org/10.1007/978-3-030-71771-1_13

[1–16]. In [1–5] methods for calculating the location of hubs are considered to minimize the economic costs and environmental impact of transporting goods from the manufacturer to the transshipment point. The problem belongs to a mixed integer programming problem, but due to the complexity of the problem being solved, the authors propose among others a heuristic approach. The paper [6] presents methods of rational distribution of cargo flows in the transport system of Poland. The search for solutions is related to the deterioration of the environmental situation and is based on the rational distribution of cargo flows between modes of transport, taking into account various criteria. The author of the work [10] suggests to pay attention to the principles of green logistics when organizing the distribution of cargo flows in the transport system. Articles [7, 9, 12, 15] study issues related to the influence of the volume of cargo flows and the capacity of transport connections on the development of ports. The authors of the articles [8, 11, 14] suggest paying attention to the regionalization of ports in order to plan their development and organize effective transportation of goods from the interior to the ports. In the papers [13, 16], a cluster approach to research the functioning of seaports in a competitive environment is considered.

This article is devoted to the development of General approaches to the accumulation and distribution of export cargo flows in the port transport and technological system, taking into account the interests of production, transport and warehouse business structures when working in a competitive transport market.

It follows from the above that modern multimodal transport requires the development of existing methods for finding rational distributions of port cargo and wagon flows, as well as the development of new methods for organizing and interacting freight transport participants. A natural component of research is to compare these methods in terms of their effectiveness and the coincidence of the results obtained.

2 Determination of the Most Suitable Cargo Loading Areas for Transportation to the Port Based on the Modified Gravitational Method

A large number of grain loading stations, closely located seaports, branching and saturation of communication lines, as well as the presence of sections of international transport corridors identified the North-Caucasus Railway (NCR) as the object of research. In this chapter, for the efficient organization of multimodal transportation of grain cargo, it is proposed to single out an element of the transport and technological system—the port cluster.

A port cluster is a combination of individual infrastructure elements: storages, transport routes, and other elements, the use of which when organizing grain transportation to the port gives the best economic result.

The purpose of forming a port cluster is:

1. Creation of a single information space between the storages and transport infrastructure.
2. Identification of the most effective cargo transportation schemes, taking into account the geographical location and economic parameters of transportation.
3. Improving the rhythm of transportation due to redistribution or the establishment of a sign of the lack of priority of cargo flows following from other clusters.

When a port cluster is formed, "border" zones can be formed—zones of influence of several ports at the same time. Or "islands"—clusters that gravitate to a specific port but are not included in the main cluster structure.

Let us analyze the clustering possibilities of the ports of the Azov-Black Sea basin and objects of transport and technological infrastructure based on the modified Sheffle—Weber gravitational method [16].

To characterize the "gravity" bonds of grain storage railway stations and ports, we use the fractional power function of the gravitational model

$$W_{ij} = \alpha \frac{\sum_{1}^{i} E_i \sum_{1}^{j} E_j}{l_{ij}^2},$$

where W_{ij}—the value of the traffic flow between the objects of the transport and technological system; l_{ij}—distance between i č j grain storage railway stations (including ports); E_i, E_j—capacity of the i and j grain storages railway stations; α—proportionality coefficient, which is determined by the formula $\alpha = E_{min}/E_{max}$.

The larger the magnitude of the gravitational bonds 1, the more stable the transport links between loading and port grain storage railway stations.

Based on the matrixes of distances between loading stations and ports, the characteristics of storages, using the gravitational method, the calculation of indicators between loading stations and ports was performed.

In total, more than 100 grain storages located at the NCR training ground and having railway access roads were investigated. An example of calculating the gravity bonds for individual stations of is presented in Table 1.

Table 1 The results of the calculation of the gravity bonds between loading stations and ports stations

Loading stations/ Port stations	**St1**	**St2**	**St3**	**St4**	**St5**	**St6**	**St7**
P1	0,143	0,05	0,12	1,49	0,152	0,164	0,161
P2	0,07	0,031	0,06	0,42	0,078	0,089	0,078
P3	0,035	0,018	0,029	0,014	0,181	0,048	0,038
P4	0,032	0,021	0,018	0,058	0,207	0,088	0,068
P5	0,092	0,055	0,045	0,17	0,76	0,27	0,229
P6	0,017	0,011	0,007	0,029	0,022	0,044	0,034

As a result of calculations of gravity indicators for all objects of the research sample, port attraction zones—port clusters were obtained: **Đ1, Đ2, Đ3, P4, P5, P6.**

The presented clustering allows you to determine the boundaries influence of the ports, based on the geographical features of their location. The formation of through freight trains within port clusters seems to be the most rational for the corresponding port in terms of its location with respect to loading points.

As a result of calculating the gravity bonds between loading stations and ports, it was found that transportation will be effective between **St1—P1; St2—P5; St3—P1; St4—P1, P2; St5—P3, P5; St6—P4, P5, P6; St7—P5.**

3 Analysis of Regional Cost Indicators of Freight Traffic

Let's move on to the first stage of constructing a mathematical model of the territorial oligopolistic freight transportation market by the method of economic and geographical delimitation of the "influence areas" of loading stations. At this stage, the necessary specific cost indicators of the transportation process in this region are identified and their preliminary analysis is carried out for the existence of possible patterns. For the seven loading stations and six port stations considered in this work, the corresponding statistical data are given in Table 2.

After analyzing the numerical data placed in Table 2 on the transportation of grain cargoes at the considered landfill, we can conclude about the linear dependence of the cost C (thousand rubles) of transportation of one wagon on the length l (km) of the route traveled. This is well illustrated by the graphic images shown in Figs. 1 and 2 for four loading stations. It should be noted that this type of dependence is very clearly manifested for transport ranges of different locations and lengths.

So, we will assume that in the mathematical model of the territorial transportation market under consideration, the expression of the dependence of the cost of C (thousand rubles) of transportation of one wagon on the length of l (km) of the route traveled by has the form:

Table 2 The statistics of grain cargo transportation

Port stations	Loading stations													
	St1		**St2**		**St3**		**St4**		**St5**		**St6**		**St7**	
	l	c	*l*	c	*l*	c	*l*	c	*l*	c	*l*	c	*l*	c
P1	207	23,2	362	30,2	211	23,2	81	17	233	24	269	26,1	199	22,2
P2	262	25,7	417	31,5	263	25,7	136	19,4	288	26,6	324	27,7	254	24,9
P3	339	29,1	494	35,3	343	29,1	213	23,2	173	21,3	401	31,5	331	29
P4	489	35,3	644	41,8	608	40,1	460	34,1	224	24	412	31,5	342	29
P5	418	31,5	573	38,9	553	37,6	389	30,2	169	21,3	341	29,1	271	25,7
P6	430	32,9	585	38,9	604	40,1	401	31,5	220	23,2	353	29,1	283	26,6

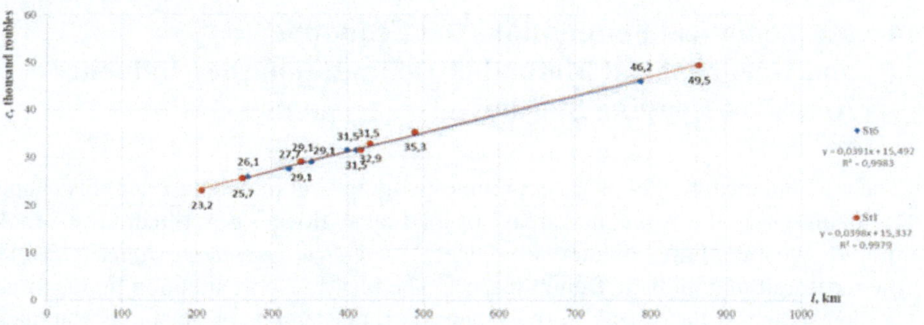

Fig. 1 Graphical depiction of freight cost dependencies for loading stations **St6** and **St1**

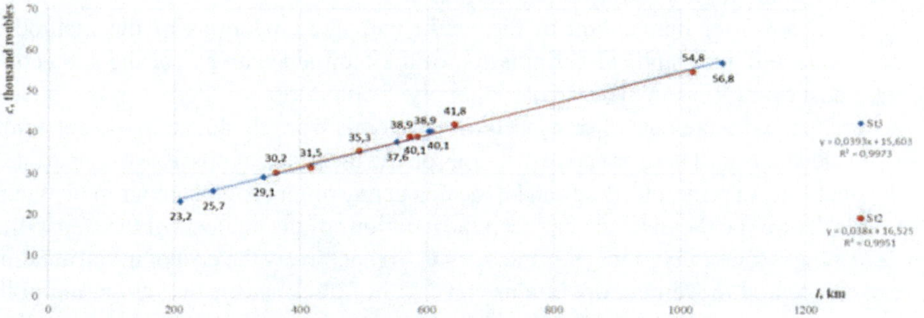

Fig. 2 Graphical depiction of freight cost dependencies for loading stations **St3** и **St2**

$$c = p + ql. \tag{1}$$

The coefficients p and q represent the costs of the initial-final operations and movement operations (per 1 km of track), which fall on one wagon. As a result of processing by the least squares method of the statistical data from Table 2 for each loading station, expressions of the form (1) are obtained, which are shown in Table 3.

The expressions given in Table 3 show that the difference between loading stations is manifested not only in the costs of initial-final operations, but also in the costs of movement operations. Note that the specified cost differentiation takes place, despite the normalized payment to the network carrier (JSC "Russian Railways"). Thus, even within the framework of the seven loading stations under consideration, the operator does not have unity in the regulators of expenses for movement operations.

Table 3 Dependences of the cost of transportation of one wagon for the loading stations

St1	St2	**St3**	**St4**
c = 15,34 + 0,039**l**	**c** = 16,52 + 0,038**l**	**c** = 15,6 + 0,039**l**	**c** = 14,27 + 0,041**l**
St5		**St6**	**St7**
c = 14,56 + 0,041**l**		**c** = 15,49 + 0,039**l**	**c** = 15,08 + 0,039**l**

4 Mathematical Foundations the Economic and Geographical Method of Delimitation the "Influence Areas" of Loading Stations

The mathematical model of the economic-geographical method used in this chapter for delimitation the "influence areas" of loading stations is described, for example, in [17]. We are talking about a very general *geometric Euclidean model* (*GEM*) of the territorial oligopolistic freight market. The real transport situation in this model is represented in the "ideal" form, when the route lengths are taken as Euclidean distances on a flat geographical map between their beginnings and ends. This assumption is quite acceptable, for example, in situations with a fairly dense transport network in the region under consideration.

Here is a brief description of the mathematical foundations of the method of economic and geographical delimitation of the "influence areas" of the subjects of the transport process in the case of a duopoly. Delimitation of these areas is made in accordance with the cost of transportation of goods from the location of each entity to the destination. There are no restrictions on the locations of possible transportation destinations. In principle, they can be located at any point on a flat geographical map.

Let L (km) be the distance between the locations of the subjects of the transportation process (duopolists), $p_i > \mathbf{0}$ and $q_i > \mathbf{0}$—respectively the cost of initial and final operations and movement operations (on 1 km of track) spent by the i th duopolist ($i = 1{,}2$) (say, per wagon). The resulting analytical expressions that define the lines we are interested in are somewhat simpler if we make the following assumption. We introduce a (conditional) Cartesian coordinate system on the plane so that the 1st and 2nd duopolists are at points $O(0,\mathbf{0})$ and $A(L,\mathbf{0})$, respectively. Then the line delimiting the "influence areas" of these duopolists is written by the equation:

$$p_1 + q_1\sqrt{x^2 + y^2} = p_2 + q_2\sqrt{(x - L)^2 + y^2} \tag{2}$$

Generally speaking, an equation of the form (2) sets an algebraic curve on the plane that has a fourth order [18]. To be more precise, this curve is a Descartes oval (the foci of which are located at points in the plane $O(0, \mathbf{0})$ and $A(L, \mathbf{0})$). Its important special case is Pascal's limaçon [19].

In order to give a clear idea of what the lines that delimit the "influence areas" of duopolists look like, as well as how the entire territorial picture of the freight transport process can change when the parameters $p_i > \mathbf{0}$ and $q_i > \mathbf{0}$ change, consider the following example. Despite the fact that specific numerical data are used, the two fourth-order algebraic curves given below have a fairly general and very characteristic appearance for our transport and logistics research.

Let's say, for example, $L = 20\mathbf{0}$, $p_1 = 6{,}1\mathbf{6}$, $p_2 = 20{,}1\mathbf{9}$, $q_1 = 0{,}0\mathbf{9}$ and $q_2 = 0{,}08$. Then the line that delimits the "influence areas" of duopolists is Pascal's limaçon (Fig. 3). The inside of limaçon (shaded area) is the "influence area" of the 1st duopolist and the outer part appearance of limaçon (that is, the rest of the plane lying outside the line)—"influence area" of the 2nd duopolist. It turns out that since the cost of

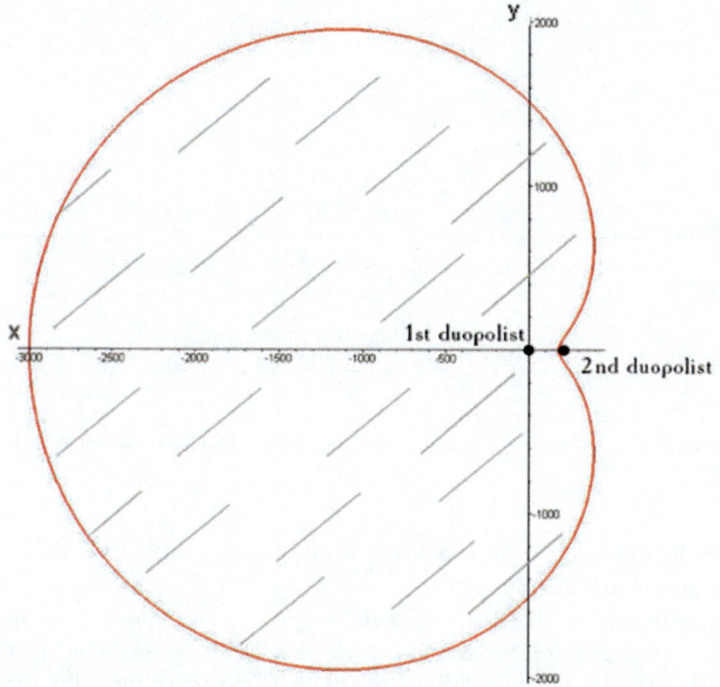

Fig. 3 The line that limits the "influence areas" of duopolists in the case when $q_1 = $ **0,09** and $q_2 = $ **0,08**

movement operations for the 2nd duopolist is less than for the 1st (note that only by 0,01), the 2nd duopolist is more competitive when transporting to many destinations that are closer to the 1st. (We are talking about points located far enough away from the 1st duopolist to the west.) The "influence area" of the 1st duopolist is nevertheless a rather large set for the reason that its costs of initial-final operations are substantially less than that of the 2nd duopolist.

We will show how the duopolistic territorial picture of the freight transport process will change (and dramatically) if we increase (just a little) the expenses of the 2nd duopolist for movement operations.

Now let's say, for example, $q_2 = $ **0,1**. Then the line that limits the "influence areas" of the duopolists in the previous case will stretch and then go into a Descartes oval, which will limit the "influence areas" of the 2nd duopolist (Fig. 4). (These metamorphoses of line are very clearly illustrated by the graphical and animated features provided by a computer mathematics system.) Now the "influence area" of the 2nd duopolist will be the bounded set on the plane. Since the costs of both initial and final operations for the 2nd duopolist are now greater than for the 1st, its competitiveness will be limited to a very small "size" (compared to what was previously) area on the plane indicated by crosses (Fig. 4).

Here you should pay attention to the relevance of the use of information technologies in research. Performing the necessary algebraic transformations, as well as the

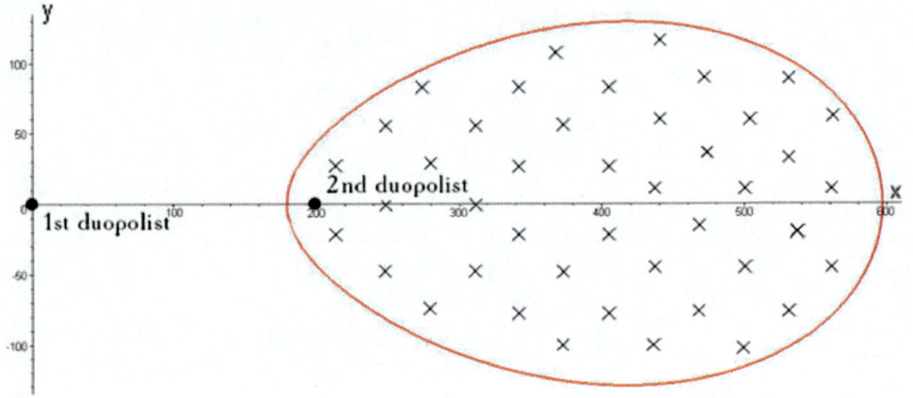

Fig. 4 The line that limits the "influence areas" of duopolists in the case when $q_1 = 0{,}09$ and $q_2 = 0{,}1$

corresponding geometric constructions, would be very difficult without the help of a computer mathematics system.

Let's return to the general duopolistic situation. We denote $p = p_2 - p_1$ and for certainty we assume further that $p \geq \mathbf{0}$. (The case where $p < \mathbf{0}$ is treated similarly.) We will get rid of irrationalities in Eq. (2) by performing the corresponding transformations in the environment of the computer mathematics system. We get the equation of line which delimits the "influence areas" of two duopolists, which has the following form:

$$\begin{aligned} &Q^2x^4 + 4q_2^2LQx^3 + 2Q^2x^2y^2 - \left(4p^2q_2^2 + 2\left(p^2 + q_2^2L^2\right)Q - 4q_2^2L^2\right)x^2 \\ &\quad + 4q_2^2LQxy^2 + \left(8p^2q_2^2L - 4\left(p^2 + q_2^2L^2\right)q_2^2L\right)x + Q^2y^4 \\ &\quad - \left(4p^2q_2^2 + 2\left(p^2 + q_2^2L^2\right)Q\right)y^2 - 4p^2q_2^2L^2 + \left(p^2 + q_2^2L^2\right)^2 = 0. \end{aligned} \tag{3}$$

(For brevity of the record we have entered the notation:.)

From Eq. (3) it can be seen that the line under consideration has a fourth order (and, for both variables x and y) if and only if $Q \neq \mathbf{0}$. Let's pay attention to the following interesting fact from a mathematical point of view. This line cannot have a third order, since at $Q = \mathbf{0}$, its order jumps downward and a line is obtained no higher than the second order. In previous works, the authors have shown that (if we exclude trivial cases) it is either a circle or a branch of a hyperbola. At the same time, it should be noted that it is not obvious (and thus non-trivial) from both the mathematical and transport and logistics point of view that the specified line cannot be any other second-order curve. (That is, it can't be an ellipse proper, or a parabola.)

5 Building the *GEM* of the Oligopolistic Freight Market

Let's proceed to the direct construction of a mathematical model of the oligopolistic freight transport market, for which the term *geometric Euclidean model* (*GEM*) was introduced in the previous works of the authors (see, for example, [17]). As noted above, the resulting territorial picture of the freight transport market is a priori conditional. In order not to increase the volume of this article with additional constructions (which are hypothetically possible), we will not discuss the *geometric routing model* (*GMM*) here. The fact is that the transport polygon in question has a fairly dense network of Railways. We refer the reader interested in the *GMM* to the article [17].

So, based on the analytical expressions of transportation costs shown in Table 3, we will get a territorial picture of the oligopolistic transportation market in the region within the framework of the company under consideration. The loading stations can be positioned as participants in an oligopoly, between which one of the types of intra-company competition is implemented. (That is, there is a non-antagonistic competition for the most efficient use of the company's resources.)

Recall that the indicator that determines the delimitation between the "influence areas" of loading stations in the *GEM* is the cost of transporting goods from these stations to their destinations. In the previous section, we discussed in detail the duopolistic situation in which the "influence areas" of two loading stations are separated by a line representing a certain Descartes oval. It follows from this that in the case of an oligopoly the boundary of the "influence area" of any of the seven loading stations under consideration will consist, generally speaking, of six components, each of which is some part of Descartes' oval. In a duopolistic situation, this oval delimits the station in question with any one loading station out of six others. (These components may be unconnected sets on the plane.) Note also that the complete territorial picture of the oligopolistic market created by seven loading stations is described, generally speaking, with lines.

On a flat geographical map of the region under consideration, we will introduce a Cartesian coordinate system, the beginning of which is the location of the station **St1**. The axes of the abscissus and ordinate are located in the direction from west to east and from south to north, respectively. This "organizational" geometric construction is performed for conditional "binding" of the market model under construction to the real territory. For Fig. 5 shows the territorial picture obtained by the computer mathematics system (in this case, the Maple system) of the oligopolistic grain transportation market formed by the seven loading stations considered in the article. The system performed this construction based on Eq. (2) and the transport cost expressions shown in Table 3.

We now give a few comments about the results obtained in this section. All six considered port stations fell into the "influence area" of only two loading stations. Namely, the port stations **P1**, **P2** and **P3** fell into the "influence area" of station **St4**, and the port stations **P4**, **P5** and **P6** fell into the "influence area" of station **St5**. (Such a result would be quite expected, if we proceed only from the territorial location of the entire set of stations under consideration.) We draw attention to the

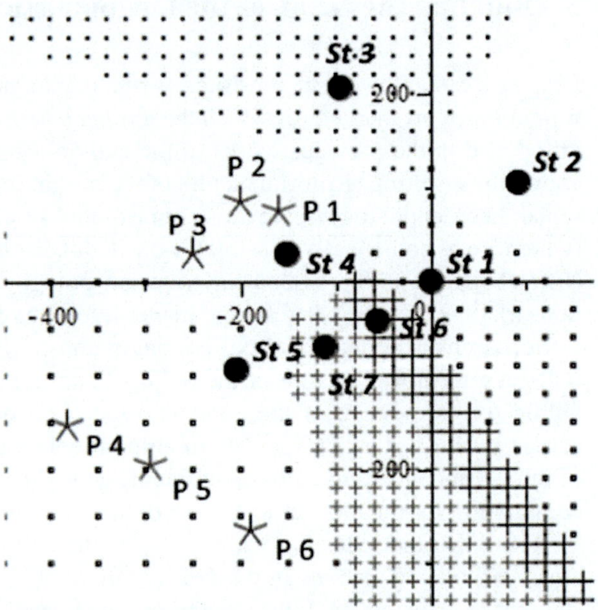

Fig. 5 Territorial picture of the oligopolistic grain transportation market, obtained on the basis of *GEM*

fact that in this situation, due to insignificant differences in the cost of initial and final operations and the cost of movement operations for all loading stations, the cost factor did not work (which, generally speaking, can significantly affect the territorial picture of transportation). Thus, the loading stations, which are farther away from the port stations than stations **St4** and **St5**, failed to "capture" any of the port stations in own "influence area".

It should also be noted that the results obtained by the economic-geographical method of separating the "influence areas" of loading stations, as a whole, turned out to be quite coincidental with the corresponding results obtained in Sect. 2 on the basis of the modified gravitational method. Namely, as a result of applying the modified gravitational method, the greatest attraction of the **St5** loading station to the port stations **P4** and **P5** is shown, and the loading station **St4** to **P1**, **P2**, **P3**. Due to the fact that this method does not provide for accounting for the cost of initial-final and mobility operations, in the resulting model, none of the loading stations can be unclaimed in the territorial picture of the transportation process.

6 Possible Adjustments to the *GEM* in the Operational Situation

Here we give an idea of some of the possibilities presented by the *GEM* and the computer mathematics system for adjusting the territorial picture of the oligopolistic market from the point of view of rationalizing the transportation in the current operational situation.

In Sect. 4, we have already seen how the territorial pattern of transportation changes in the case of a duopoly, even with a very slight change in the cost of movement of one of the subjects of the transportation process. Let us show now with two examples how the oligopolistic territorial picture of the process of freight transportation changes when the cost of the initial-final operations of any oligopolist changes. In Fig. 6 shows the corresponding picture, which is obtained if the cost of the initial-final operations performed at the loading station **St3** is reduced from $p = 15, \mathbf{6}$ to $p = 10, \mathbf{7}$.

Now, only one port station **P3** falls into the "influence area" of the **St4** loading station. Stations **P1** and **P2** are in the "influence area" of the **St3** loading station. We also note that the "influence area" of **St4** station is now a bounded set on the plane, and having a very small area. These changes are, of course, expected. However, *GEM* allows to evaluate them in an exact quantitative sense.

We will also give a metamorphosis that will happen to the oligopolistic freight market if we reduce the cost of the initial and final operations at the **St7** loading station. In Fig. 7 shows the corresponding territorial picture if the indicated cost is reduced from a value of $p =$ 15,**08** to $p =$ 13,**08**.

With this change, the two port stations **P4** and **P5** will remain in the "influence area" of the loading station **St5**. The "influence area" of the loading station **St7** will

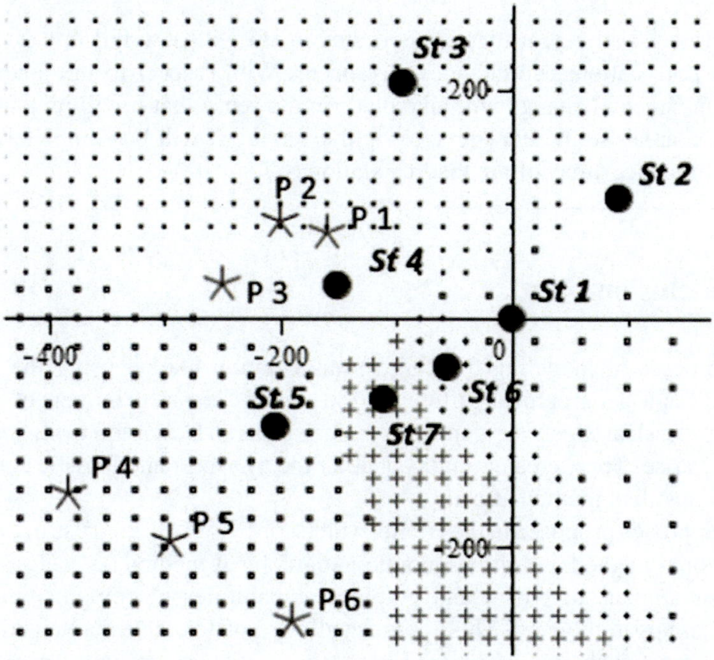

Fig. 6 The territorial picture obtained by reducing cost of initial operations at the station **St3**

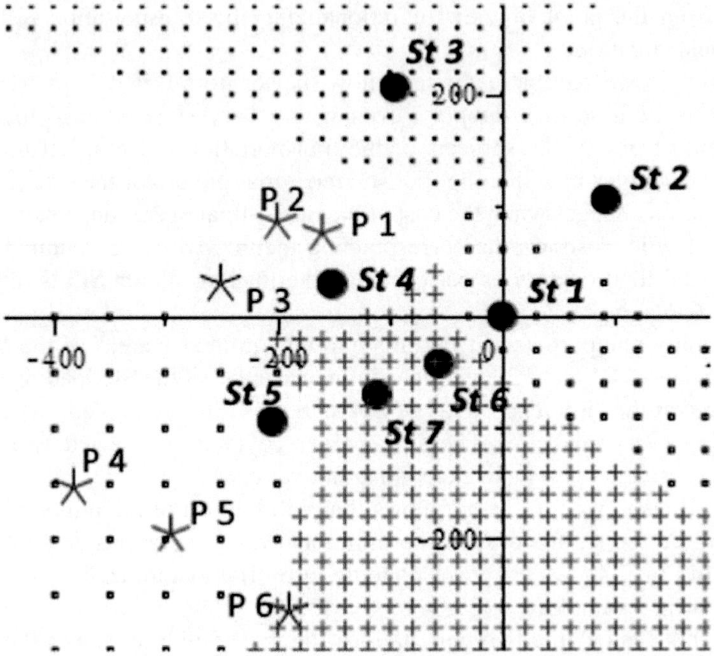

Fig. 7 The territorial picture obtained by reducing cost of initial operations at the station **St7**

expand to such an extent that the port station **P6** will now fall into it. In relation to other port stations, no changes will occur. With respect to the loading station **St6**, a fundamental change will take place in the sense that it will remain without a "influence area" at all. Her previous "influence area" will be completely absorbed by the "influence area" of the loading station **St7**.

7 Conclusion

In the process of modeling the cargo transportation market using the method of economic and geographical differentiation of the "areas of influence" of loading stations, classical algebraic curves of higher orders (Descartes ovals and Pascal's limaçon) arise. Thus, created in this chapter the transport and logistic constructions are quite justified mathematically.

The approach proposed in the article, which consists in the joint use of the modified gravitational method and the economic-geographical method, as well as in the use of computational, analytical and graphical capabilities of computer mathematics systems, is apparently new. This approach allows you to create mathematical models of the cargo transportation process based on a comprehensive assessment of the transport and technological infrastructure of the railway polygon and the cost of transport services.

The method of analytical modeling of cargo transportation process management developed in the article can be supplemented with software tools adapted for the user. This makes it possible, including in the operational mode, to adjust the territorial picture of the transportation process depending on the current situation and find the most rational options for using rolling stock and infrastructure when transporting goods to the destination stations.

Acknowledgements The reported study was funded by RFBR, Sirius University of Science and Technology, JSC Russian Railways and Educational Fund "Talent and success", project number 20-38-51014.

References

1. Kwon S, Park K, Lee C, Kim SS, Kim HJ, Liang Z (2006) Supply chain network design and transshipment hub location for third party logistics providers. In: Gavrilova M et al (eds) Computational science and its applications—ICCSA 2006. Lecture notes in computer science, vol 3982. Springer, Berlin, Heidelberg
2. Macioszek E, Sierpiński G, Staniek M (2017) Analysis of trends in development of freight transport logistics using the example of Silesian Province (Poland)—a case study. Transp Res Procedia 27:388–395
3. Kush Y, Skrypin V, Galkin A, Dolia K, Tkachenko I, Davidich N (2018) Regularities of change of the supply chain operation efficiency depending on the parameters of the transport process. Transp Res Procedia 30:216–225
4. Kush Y, Tonkoshkur M, Vakulenko K, Ryabev A, Davidich N, Galkin A (2020) The efficiency of food supply chain engineering (case study in Ukraine). Transport 55(3):51–71
5. Sierpiński G, Macioszek E (2020) Equalising the levels of electromobility implementation in cities. In: Mikulski J (ed) Research and the future of telematics, communications in computer and information science, vol 1289. Springer, Heidelberg, pp 165–176
6. Jacyna M (2013) Cargo flow distribution on the transportation network of the national logistic system. Int J Logist Syst Manag 15(2–3):197–218
7. Paulauskas V, Lukauskas V (2003) Cargo flow and transit impacts on port development. Transport 18(3):120–123
8. Notteboom TE, Rodrigue J-P (2005) Port regionalization: towards a new phase in port development. Maritime Policy Manage 32:297–313
9. Notteboom TE, Winkelmans W (2001) Structural changes in logistics: how will port authorities face the challenge? Maritime Policy Manage 28:71–89
10. Roop S (2006) The freight shuttle: the crisis in freight transportation and the opportunity for a green alternative. Paper presented at the METRANS national urban freight conference, Long Beach, 1–3 February
11. Fu Z, Li Y, Lim A, Rodrigues B (2007) Port space allocation with a time dimension. J Oper Res Soc 58:797–807
12. Wiradanti B, Pettit S, Potter A, Abouarghoub W (2018) Ports, peripherality and concentration—deconcentration factors: a review, Maritime Bus Rev. https://doi.org/10.1108/mabr-09-2018-0040
13. Nežerenko O, Koppel O, Tuisk T (2015) Cluster approach in organization of transportation in the Baltic Sea Region. Transport 32(2):1–13

14. Valls J, Langen P, Garcia-Alonso L, Vallejo-Pinto J (2020) Understanding port choice determinants and Port Hinterlands: findings from an empirical analysis of Spain. Maritime Econ Logist 22:53–67
15. Guerrero D (2018) Impacts of transport connections on port hinterlands. Reg Stud 53(4):540–549
16. Chislov O, Bogachev T, Kravets A, Bogachev V, Zadorozhniy V, Egorova I (2020) Time parameters optimization of the export grain traffic in the port railway transport technology system. In: Sierpiński G (ed) Smart and green solutions for transport systems. TSTP 2019. Advances in intelligent systems and computing, vol 1091. Springer, Cham, pp 126–137
17. Chislov ON, Bogachev VA, Zadorozhniy VM, Demchenko OI, Khan VV, Bogachev TV (2019) Modelling of the rail freight traffic by the method of economic-geographical delimitation in the region of the South-Easter Coast of the Baltic Sea. Transp Probl 14(2):77–89
18. Vinogradov IM (1979) Mathematical Encyclopedia, vol 2. Soviet Encyclopedia, pp 79–80
19. Vinogradov IM (1984) Mathematical Encyclopedia, vol 4. Soviet Encyclopedia, pp 226–227

Evaluation of Global Transportation Systems—Classical Approach Versus Intelligent Approach. Case Study Analysis

Barbara Galińska

1 Introduction

Constant changes in economy as well as an increasing market competition have a significant influence on procurement process in manufacturing companies. Such companies aim at minimization of the production costs and time of products and information flow, whereas remaining the high-level quality of the goods and customer service. This can be achieved by cooperation with the suppliers located in developing countries where manufacturing costs are much lower in comparison to other regions. China is therefore the country, which plays a significant role in migrating production. Nevertheless, its location is unfavourable for European manufacturers because of the far distance, high delivery costs and long delivery time. All these aspects contribute to facilitation of importing materials from China by adjusting the most efficient transportation systems. Thus, the detailed analysis of goods delivered, their qualities and characteristic, are the factors, which mostly determine the best transportation variant. The evaluation, which includes number of criteria, must be carried out in a logical manner based on the intelligent approach. This chapter presents the process of evaluation and ranking of global transportation systems that issue affects a whole management of global supply chains. The author refers to the actual case study, which analyses global transportation systems between China and Central Europe (Poland). Presented results are based on two approaches—classical, involving scoring method and on the intelligent approach, which implies multi-criteria analysis, in particular Promethee II method.

Also, both of them represent alternative, multi-modal transportation solutions that constitute the considered variants of delivering goods. The transportation systems are

B. Galińska (✉)
Faculty of Management and Production Engineering, Lodz University of Technology, Lodz, Poland
e-mail: barbara.galinska@p.lodz.pl

G. Sierpiński and E. Macioszek (eds.), *Decision Support Methods in Modern Transportation Systems and Networks*, Lecture Notes in Networks and Systems 208,
https://doi.org/10.1007/978-3-030-71771-1_14

constructed heuristically, based on the author's intuition and the expert knowledge. They might involve road, railway, sea and air transportation. Since China is a very meaningful, global supplier that delivers raw materials, components and finished goods on a large scale, worldwide, the analysis of the Chinese market makes, in the author's opinion, the presented results quite interesting.

Yet, the overall research goal of this chapter is to evaluate the global transportation systems and select the most desirable ones. Such evaluation (including a set of evaluation criteria) has been carried out with the use of two different methods i.e. scoring method (commonly used in the classical approach to decision making) and Promethee II method (used in the intelligent approach).

In the course of the computational experiments, the author generates final rankings of transportation options. Then, the rankings, generated by different computational algorithms, lead to final conclusions. Finally, the best option of the transportation system is recommended. The originality of this work consists of the description and confrontation of all components of analysis of transportation's variants, based on the different approaches.

The chapter is composed of 5 sections. The first one is a short introduction to the problem, defining the aim of the study. The second section includes the description of the methodological background of the research. It presents the analysis of the global transportation systems, description of the transportation options between China and Central Europe (mainly considering Poland) and finally characteristic of the research methods i.e. scoring method and Promethee II. Section 3 is focused on decision situation, it mainly presents characteristic of the global transportation systems and their evaluation criteria. The fourth section includes the results of computational experiments generated with the applied methods. In Sect. 5 the final conclusions are presented. The chapter is supplemented by a list of references.

2 Methodological Background of the Research

2.1 Global Transportation System

Transportation system is the main link between the customers, warehouses, suppliers, manufacturing companies and other distribution channels. All of these above mentioned components have a dominant position in logistic system, which sometimes results in a duty suspension movement at the certain levels. Only transport ensures connection between all these elements. It can be therefore concluded that transport creates so called bridge between purchasers and traders, filling the gap in the delivery chain [1–7].

Transportation systems can be classified differently based on different characteristics/criteria [8–15]:

- scope and range,
- modal diversity,

- category of medium it carries/moves,
- environment the system operates in.

Above mentioned classification of criteria let us distinguish international/global, multi-modal, transportation systems operating in various environments (sea, air, surface) that are the object of the research presented in this chapter. Such systems may cover a specific group of countries, one or several continents or even the whole world and involve the combination of various transportation modes such as: rail, road, sea and air transportation [2, 16]. They have a huge impact on the international trade exchange where costs and time of transportation are very important [17].

These days, it is especially necessary to look for solutions, which will increase competitive advantage of one company over another. Minimizing the prices of the goods by means of lowering the manufacturing costs is one of the way to achieve it [18]. It is often inevitable to transfer manufacturing to another countries or acquire materials from suppliers who come from countries where the labour force is low-priced [19]. The example can be China, where number of goods imported from this country is constantly increasing on the European market.

2.2 *Transport Capacity Between China and Central Europe (Poland in Particular)*

China maintains a leading position in the procurement market in the world. Its dynamic development provides favourable conditions for foreign investments (mainly in processing) and transfer of new technologies. Its main advantage is low labour cost, however, there are also many other important aspects [20]:

- low salary assignment connected with indirect costs,
- underestimated conversions of Chinese Yuan (CNY), affecting the growth of competitiveness in international trade,
- growth of productivity and labour,
- entrepreneurship and discipline—the features of the Chinese employees, highly desired on the market.

China is one of the most significant trading partner for the European Union, whereas European Union for the China acts as the major and the main one. In 2014 the total value of trade between these two regions was EUR 467 billion, which means transport of goods, involving various modes of transport, valued over one billion euro on everyday basis. These amounts are clearly increasing by a few per cent every year.

Taking into account all the European countries, Poland has not only the biggest market in the Eastern Europe but also the nearest location. Thus, the number of Chinese companies which supply Europe, involving Polish cooperation, is constantly growing. Poland also stands out in terms of transport potential, widely used in the distribution of Chinese goods to Europe.

Table 1 International trade of goods between Poland and China by means of sea transport

Time period	Import (thousand tonnes)	Export (thousand tonnes)
2014	1243,7	358,7
2015	1080,2	331,8
2016	1251,7	414,7
2017	1713,6	774,0

Source Self—study based on data provided by Central Statistical Office in Poland

Main transport branch for trading goods from China is transport by sea. However, such solution implies a lot of obstacles for the companies, mainly in the area of supplies planning and placing orders. The reason for that is a long transport period, remaining over 30 days [21]. Nevertheless, considering other transportation solutions and their costs in general, the trading in goods by sea transport is constantly increasing (Table 1).

The alternative for transport by sea is rail transport, providing lower transport period, of around 14–21 days. It is especially effective when it comes to urgent deliveries or production delays. Despite the reasonably high costs, in comparison to transport by sea, the number of consignments provided by rail transport is also growing (Table 2).

Other alternative is air transport, which offers the shortest delivery time, around 2–3 days. However, for the most companies this solution is too expensive, giving costs several times higher in comparison to transport by sea or to rail transport.

In order to achieve the satisfactory value for the customer, it is necessary to manage the efficient flow of materials. That is the main reason for a dynamic development of supply chain, including the global ones. Yet, this generates a number of obstacles which result from usage of global transportation systems. As they vary a lot, entrepreneurs are forced to choose the most beneficial variants. Also, this was the main reason to evaluate the alternative transportation systems in the case study described below.

Table 2 International trade of goods between Poland and China by means of rail transport

Time period	Import (thousand tonnes)	Export (thousand tonnes)
2014	no data	724,0
2015	1324,0	1022,0
2016	1421,0	1075,0
2017	1672,0	1225,0

Source Self—study based on data provided by Central Statistical Office in Poland

2.3 *Characteristic of Research Methods (Scoring Method + Promethee II)*

The classical approach uses **the scoring method**, which is based on a score-based assessment including quantitative and qualitative features (e.g. price) of the analysed variant (global transportation system—in the case study below).

In this method, the number of scores represents the degree complying with the particular evaluation criteria. Then, scores are summed including their criteria weights. In order to perform the scoring method, five stages must be taken into consideration, i.e.:

I—naming evaluation criteria including criteria parameters, which are all basis for the quantification process,

II—establishing point scale for the criteria and their parameters,

III—naming the weights for criteria and their parameters in order to identify their value for the DM,

IV—adding up the sum of scores,

V—selection of the final variant.

The results in scoring method can be represented by a graphical form as it improves the clarity of the general idea.

On the other hand, intelligent approach involves more sophisticated research methods including decision making, especially the selection of the most desired variant. These methods are in particular: Multiple Criteria Methodology, Group Decision Making, Artificial Neural Networks, Metaheuristic, Fuzzy Logic, Case—Based Reasoning, Expert Systems, which significantly rise the rationality of the decision made [22].

One of the tools for intelligent decision making is a Multiple Criteria Decision Making (MCDM) also known as a Multicriteria Decision Aid (MCDA) or Multicriteria Decision Support [23, 24]. It is focused on the development and implementation of decision support tools, rules and methods to confront complex decision problems involving multiple criteria, goals or objectives of conflicting nature [25]. The methodology of MCDM/A has a universal character and can be applied in various cases when a DM solves a so called multiple criteria decision problem (MCDP) [26]. To solve this problem multiple methods can be applied. The differences between these methodologies involve the form of the models, the model development process and theirs cope of application [27]. One of the most popular method implementing the outranking relations framework is **the Promethee method** [28, 29]. It was introduced to preference rank a set of decision alternatives, based on their values over a number of different criteria. Put simply, a ranking of alternatives is established, based on the accumulative preference comparisons of pairs of alternatives' values over the different criteria (using generalised preference functions) [30]. The Promethee method will provide the DM with a ranking of actions (choices or alternatives) based on preference degrees. The method falls into three main steps [31]:

I—the computation of preference degrees for every ordered pair of actions on each criterion,

II—the computation of unicriterion flows,

III—the computation of global flows.

Finally, both above mentioned methods (scoring method and Promethee II) will be further applied in the practical part of this chapter in order to select the most desired variant—global transportation system—in the decision situation.

3 Description of Decision Situation

3.1 Verbal Description

The aim of the chapter is to evaluate and rank global transportation systems for the company operating in household industry. The entity is a part of the international corporation, which is the leading producer of such appliances in the world.

In accordance with the company policy some of the components are delivered from the Far East. The biggest share constitute materials purchased from the supplier whose manufacture is located in China. It is one type of material which stands for five different goods (which differ in specification and technical parameters). These components are used in about 67% of manufactured appliances, one item per one appliance.

What is more, the process of materials delivery from China is connected with some obstacles caused by the distance what results in high delivery time. It influences the quantity of materials stored in the warehouses. It is also the main reason why the Plant Director of the company, acting as DM in the decision situation described, is determined to rationalize the supply chain of purchased goods in order to minimize the goods in store.

One of the ways to achieve it, is by rationalizing the supply chain of purchased components so that it would be possible to reduce the costs and time of delivery as well as to improve such parameters as delivery safety and delivery timeliness.

3.2 Characteristics of Variants—Global Transportation Systems

To date, the analysed company used mainly one transportation system, the combination of transport by sea and road-rail transport (above mentioned variant V1). The Plant Director, acting as DM, decided to carry out an analysis of five other transportation options, mainly taking into consideration the time of transport, its costs and timeliness, so that finding the most favourable variant would be possible. Considered variants cover the distance from China/Qingdao (sea port), Chengdu (railway station) or Qingdao-Liuting (airport) to Poland/Lodz (the company's headquarter)V1–V6 (Table 3).

In consistency with Incoterms 2010 all the above mentioned variants entail the supplier' responsibility for delivering goods from his factory to the indicated place, i.e. Qingdao sea port, Chengdu railway station or Qingdao-Liuting Airport. Thus, all transportation systems cover the distance from the collection point in China to company's headquarter in Lodz (Poland).

Table 3 Variants—global transportation systems—verbal description of decision problem

Variant	Type	Verbal description
V1	Sea transport + sea transport + road—rail transport	sea transport from port in Qingdao to Hamburg port sea transport from Hamburg to Gdynia port rail transport from Gdynia to container terminal in Stryków road transport from Stryków to the company's headquarter in Lodz
V2	Sea transport + sea transport + road transport	sea transport from Qingdao to Hamburg port sea transport from Hamburg to Gdynia port road transport from Gdynia to the company's headquarter in Lodz
V3	Sea transport + road transport	sea transport from Qingdao to Hamburg port road transport from Hamburg to company's headquarter in Lodz
V4	Rail transport + road transport	rail transport from Chengdu to Lodz railway station road transport from Lodz railway station to the company's headquarter in Lodz
V5	Rail transport + rail transport	rail transport from Chengdu to reloading terminal in Lodz rail transport from Lodz railway station to company's headquarter in Lodz (the company owns its siding)
V6	Air transport + road transport	air transport from Qingdao-Liuting Airport to Warsaw Airport road transport from Warsaw Airport to the company's headquarter in Lodz

Source Self—study

3.3 Variants Evaluation Criteria

As the DM has used only one transportation system so far (above mentioned variant V1), the selection of the new system ought to be well-thought and based on the thorough analysis. The evaluation's criteria have been constructed according to the model proposed by B. Galińska and J. Żak, which distinguishes seven main evaluation criteria. The importance wages of the criteria were formulated on the basis of the interview with the DM, his preferences and aspirations. Due to research limitations, criteria K1–K7 and their sub-crietria have been only enumerated, without any detailed descriptions (Table 4).

Based on the above mentioned definitions of criteria the appropriate data have been collected and the evaluation matrix has been constructed (Table 5). The matrix includes evaluations of all variants (global transportation systems V1–V6) on all criteria and sub-criteria.

The aim of computational experiments was to indicate the most adequate global transportation system in described case study. The results, based on scoring method and Promethee II, are presented in the next section of this chapter.

4 The Results of Computational Experiments

The results of the computational experiments are limited to presentation of the final rankings, identifying the positions of particular global transportation systems (variants V1–V6) toward each other.

4.1 Ranking of the Variants with Application of Scoring Method

In accordance with the scoring method algorithm, the evaluation criteria of global transportation system have been defined. Then, the scoring criteria and their parameters have been determined. Next stage required naming the criteria weights, so that their total amount is 100% (Table 4) and defining scoring criteria parameters by entering a score for each criterion and multiplying each score against the weight to get a weighted score (Table 6). Finally, scoring analysis has been performed, indicating the most favourable variant (the one, with the highest score).

Table 4 Evaluation criteria in the case study

Criterion	Sub-criterion	Weight of Criterion
K1: Transportation Costs	The criterion is defined as the total/overall costs of moving goods from their origin to final destination, expressed in monetary units [PLN]	0.33
K2: Delivery Time	The criterion is defined as the total duration time of the delivery process, expressed in days	0.14
K3: Timeliness of Delivery	The criterion defines the share (percentage—wise) of on-time deliveries	0.18
K4: Reliability of a Transportation System	The criterion measures the share (percentage—wise) of damaged shipments in the delivery process	0.02
K5: Flexibility of a Transportation System	K5.1—frequency of delivery (defined as a number of potential deliveries in a monthly period)	0.01
	K5.2—minimal size of the shipment (expressed on the 1 to 10 point scale)	0.01
	K5.3—capacity availability (average share of container's capacity evaluable)	0.10
K6: Safety of a Transportation System	K6.1—number of handling operations	0.08
	K6.2—total period of the storage in the transhipment terminals (expressed in days)	0.03
K7: Customer's Comfort	K7.1—ability to monitor the shipments and its status information (expressed in point)	0.04
	K7.2—additional services, including the duration of free of charge storage of shipment	0.04
	K7.3—payment periods (expressed in days)	0.02

Sourcs: Self—study

4.2 *Ranking of the Variants with Application of Promethee II Method*

Consistent with the Promethee II method algorithm, the evaluation matrix has been constructed (Table 5) and the DM's preference model has been defined in the process

Table 5 The Evaluation Matrix in described case study

Criteria		Variants					
		V1	V2	V3	V4	V5	V6
K1 [PLN]		5325	6000	6330	33210	33110	324000
K2 [DAYS]		57	52	47	20	22	7
K3 [%]		0.71	0.79	0.84	0.90	0.86	1
K4 [%]		0	0	0	0	0	0
K5	K5.1 [POINTS]	3	5	5	5	4	20
	K5.2 [POINTS]	4	6	6	4	4	8
	K5.3 [%]	0.50	1	1	0.50	0.50	0.50
K6	K6.1 [POINTS]	3	2	1	1	1	2
	K6.2 [DAYS]	11	7	7	7	10	2
K7	K7.1 [POINTS]	2	5	5	7	7	10
	K7.2 [DAYS]	10	8	7	0	5	0
	K7.3 [DAYS]	60	45	45	30	30	0

Source Self—study

Table 6 Evaluation of global transportation systems with application of scoring method

Criterion	Sub-criterion	Weight	Score						Weight*Score					
			V1	V2	V3	V4	V5	V6	V1	V2	V3	V4	V5	V6
K1		0.33	10	7	5	2	2	0	3,3	2,31	1,65	0,66	0,66	0
K2		0.14	2	3	4	8	6	10	0,28	0,42	0,56	1,12	0,84	1,4
K3		0.18	2	4	4	8	6	10	0,36	0,72	0,72	1,44	1,08	1,8
K4		0.02	10	10	10	10	10	10	0,2	0,2	0,2	0,2	0,2	0,2
K5	K5.1	0.01	2	8	8	8	4	10	0,02	0,08	0,08	0,08	0,04	0,1
	K5.2	0.01	4	6	6	4	4	8	0,04	0,06	0,06	0,04	0,04	0,08
	K5.3	0.10	5	10	10	5	5	5	0,5	1	1	0,5	0,5	0,5
K6	K6.1	0.08	2	6	10	10	10	6	0,16	0,48	0,8	0,8	0,8	0,48
	K6.2	0.03	4	6	6	6	4	10	0,12	0,18	0,18	0,18	0,12	0,3
K7	K7.1	0.04	2	5	5	7	7	10	0,08	0,2	0,2	0,28	0,28	0,4
	K7.2	0.04	10	10	8	6	8	6	0,4	0,4	0,32	0,24	0,32	0,24
	K7.3	0.02	10	7	7	5	5	0	0,2	0,14	0,14	0,1	0,1	0
Total									5,66	6,19	5,91	5,64	4,98	5,50

Source Self—study

of naming the wages of criteria and thresholds: indifference threshold q and preference threshold p. Computational experiments resulted in obtaining the final ranking and these are further represented in Fig. 1.

Final rankings obtained from the computational experiments indicated the least and the most favourable global transportation systems for the company in the case

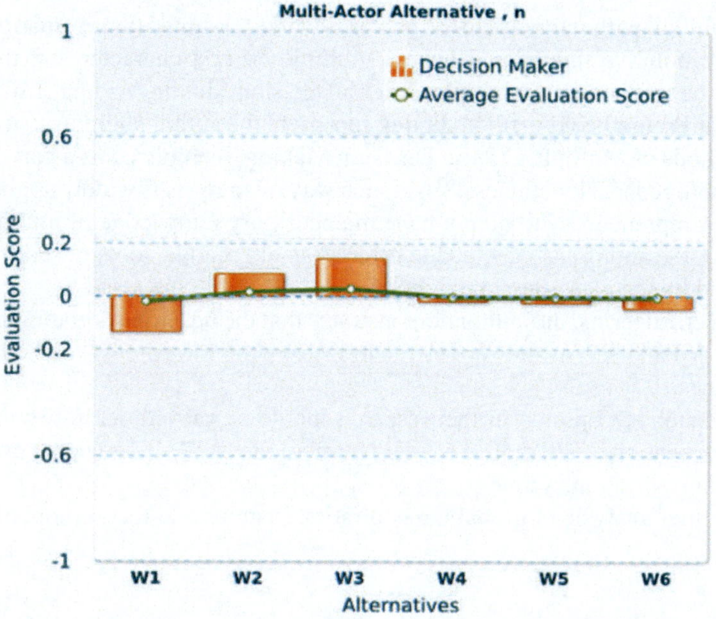

Fig. 1 Evaluation of global transportation systems with application of Promethee II method

study described. The analysis of the whole evaluation process clearly indicates variants V2 and V3, which use mostly sea transport for the majority of transportation process, as they both acquired the highest ratings. In the author's opinion, what deserves a special attention is variant V3, which ranked in the first position in Promethee II method. Although the cost of the transport (K1) in this variant is relatively higher in comparison to variants V1 and V2, the delivery time and (K2) and timeliness of delivery (K3) distinguish this variant among other. In addition, these two aspects are particularly important for the DM—Plant Director, who decided to look for alternative solutions aiming at minimizing the goods in store, by implementation of the more efficient time saving system.

5 Conclusions

The presented chapter is a comprehensive study concerning the evaluation and selection of global transportation systems for the company in industry of household appliances. It presents two different ways of selecting global transportation systems, representing two different approaches—classical approach involving scoring method and intelligent approach, which uses the principles of Multiple Criteria Decision Making/Aiding (MCDM/A) Methodology, especially Promethee II method. This research is an extension of the author' previous works [18, 22, 32, 33].

The author, with reference to the previous works, assumed that evaluation of global transportation systems should have a multiple criteria character and thus, implemented the principles of Multiple Criteria Decision Making/Aiding (MCDM/A), in particular Promethee II method. In this approach, the author attempts to use the tools and methods of Multiple Criteria Decision Making, recognized as a part of the Artificial Intelligence. Nonetheless, since such way of analysis is widely applied (mainly by the entrepreneurs who do not have the necessary knowledge of multiple criteria analysis), the author also decided to offer as an alternative, tools of scoring method, presented by the classical approach.

In practical terms, the author demonstrates that the best transportation systems are variants V3 and V2, as they are characterized by the lowest costs and other various advantages.

In the author's opinion further research should be carried out in two directions:

1. Application of alternative MCDM/A methods (ANP, UTA) to the evaluation of global transportation systems.
2. Further analysis of global transportation systems and their components.

References

1. Cavinato JL (1989) Transportation-logistics dictionary. Springer
2. Galińska B (2015) Koncepcja Global Sourcing. Teoria i Praktyka. Difin S.A
3. Pfohl HCh (2010) Logistiksysteme. Springer, Betriebswirtschaftliche Grundlagen
4. Waters D (2003) Global logistics and distribution planning. Strategies for management. Kogan Page, London
5. Macharis C (2014) Sustainable logistics. Emerald Publishing Limited
6. Sierpiński G (2013) Revision of the modal split of traffic model. In: Mikulski J (ed) Transport systems telematics 2013, vol 395. Activities of transport telematics, communications in computer and information science. Springer, Heidelberg, pp 338–345
7. Staniek M, Sierpiński G, Celiński I (2015) Shaping environmental friendly behaviour in transport of goods: new tool and education. In: 8th International conference of education, research and innovation, (ICERI2015), pp 118–123, Seville, Spain
8. Hall R (2003) Handbook of transportation science. Springer
9. Sierpiński G, Staniek M, Celiński I (2015) New methods for pro-ecological travel behavior learning. In: 8th International conference of education, research and innovation, (ICERI2015), pp 6926–6933, Seville, Spain
10. Macioszek E, Sierpiński G, Staniek M (2017) Analysis of trends in development of freight transport logistics using the example of Silesian Province (Poland)—a case study. Transp Res Procedia 27:388–395
11. Tubis A (2019) Route risk assessment for road transport companies. In: Zamojski W, Mazurkiewicz J, Sugier J, Walkowiak T, Kacprzyk J (eds) Contemporary complex systems and their dependability. Advances in intelligent systems and computing, vol 761. Springer, Cham, pp 450–463
12. Chislov ON, Bogachev VA, Zadorozhniy VM, Demchenko OI, Khan VV, Bogachev TV (2019) Modelling of the rail freight traffic by the method of economic-geographical delimitation in the region of the South-Easter Coast of the Baltic Sea. Transp Probl 14(2):77–89
13. Skupień E, Tubis A (2018) The use of linguistic variables and the FMEA analysis in risk assessment in inland navigation. TransNav—Int J Marine Navig Safety Sea Transp 12(1):143–148